Numerical Modeling of Soil Constitutive Relationship

Qingyang Ren · Jianting Zhou

Numerical Modeling of Soil Constitutive Relationship

Science Press
Beijing

Springer

Qingyang Ren
School of Civil Engineering
Chongqing Jiaotong University
Chongqing, China

Jianting Zhou
School of Civil Engineering
Chongqing Jiaotong University
Chongqing, China

ISBN 978-981-16-3230-3 ISBN 978-981-16-3231-0 (eBook)
https://doi.org/10.1007/978-981-16-3231-0

Jointly published with Science Press
The print edition is not for sale in China (Mainland). Customers from China (Mainland) please order the print book from: Science Press.

This Springer imprint is published by the registered company Springer Nature Singapore Pte Ltd.
The registered company address is: 152 Beach Road, #21-01/04 Gateway East, Singapore 189721, Singapore

Preface

The constitutive relation of soil takes the stress-strain law of soil as the research object, which is an important theoretical basis of geotechnical engineering science. At present, most of the existing elasto-plastic constitutive models of soil are based on the theory of plastic potential, and it is the only way to find the analytical expression of plastic potential. However, these models can not fully and accurately reflect the characteristics of soil, such as the impact of compression, dilatancy, especially the stress path.

In view of this, the numerical modeling of expansive soil, sand and clay is studied. Numerical modeling has many advantages over traditional modeling methods. The constitutive equation of soil is extracted directly from the triaxial test data, which overcomes the difficulty of finding the analytical expression of plastic potential. In the modeling, the initial state and loading path of soil are freely selected, so that the influence of any initial state and stress path on the constitutive relationship can be considered.

Water content is the main reason for the expansion and contraction of expansive soil and the influence of strength. It has been proved by practice that the expansive soil has water migration and transformation with the change of climate, geological environment and engineering construction activities. Therefore, under the influence of water, the properties of expansive soil change, showing different characteristics. The variability of shear strength of expansive soil is mostly related to the action of water. The anisotropic expansion caused by water absorption of expansive soil may produce deviating stress. The uneven expansion caused by different water absorption distribution may also cause the damage of expansive soil. Therefore, in engineering practice, the initial water content of expansive soil is regarded as an important criterion to predict the expansion potential.

The physical property test and Mineral chemical composition test of expansive soil are carried out. The conventional triaxial compression drainage (DCTC) and undrained (UCTC) tests with different water content and unit weight are carried out. The elastic-plastic constitutive equation of expansive soil under different water content conditions is established by numerical modeling method, and the constitutive equation is embedded in the finite element program. The deformation of the triaxial test soil sample is divided into three parts The stress-strain curve is obtained.

Compared with the corresponding test results, the two are in good agreement. At the same time, the three-dimensional surface of the stress-strain relationship in the whole stress field (P, q) under two different water contents is drawn. Through comparison, it is found that there is a significant difference between the two, which confirms that the influence of water content on the constitutive relationship of expansive soil is relatively obvious.

The cyclic loading tests of sand under the following three stress paths are carried out: equal principal stress ratio path (ertc), conventional path (DCTC) and equal P (dptc) stress path. The elastic-plastic constitutive equation of sand under three stress paths is established by numerical modeling method, and the constitutive equation is embedded in the finite element program. The accuracy of the model is verified by comparing the finite element calculation results of triaxial test soil sample deformation with the corresponding test results. The stress-strain relationship curve in the whole stress field (P, q) under three stress paths is drawn By comparing the surface and yield locus, it can be seen that there are significant differences among them, and the differences reflected by the stress-strain relation surface and yield locus have the same trend of change, which proves that the influence of stress path on the constitutive relation cannot be ignored.

The triaxial loading tests of four stress paths of clay are carried out: DCTC, dptc, UCTC and drtc. The elastic-plastic constitutive model of clay under these four stress paths is established, and the stress-strain relationship in the whole stress field is given, which is visualized as stress field Space strain surface in. Especially, the elastic-plastic constitutive model of clay under the condition of p-reduction path and undrained condition is established, which provides a practical constitutive model for soil excavation engineering and corresponding soil engineering under undrained condition. Through the visualization of stress-strain relationship and numerical simulation, it is further shown that the numerical modeling method can comprehensively describe the deformation characteristics of soil under different stress paths.

Comparing the deformation results of normal consolidated soil under four stress paths, it is found that there are significant differences in stress range, strain peak value, shape of strain surface and change trend of volume yield locus, while the shear yield locus is similar, which are caused by the correlation of stress paths. At the same time, it shows that the stress path correlation can not be ignored in the soil constitutive relationship, and the numerical modeling method is an effective method to describe the stress path correlation.

By comparing the shear and volume yield trajectories of normally consolidated soils under different stress paths, it is shown that the stress paths have a significant impact on the evolution process of strain hardening of clays. At the same time, it is not accurate to describe the stress-strain relationship of soils by assuming the fixed form of yield surface in the traditional modeling method of elastoplastic constitutive model of soils.

The results show that the numerical modeling is better than the traditional modeling method. It not only describes the dilatancy and compression of soil, but also reflects the influence of different initial states and stress paths. The numerical modeling method is feasible, effective and has a broad development prospect.

This book is mainly supported by NSFC "Research on deformation mechanism of fissured rock under equal stress ratio path (40902082)", "Research on influence of rheological characteristics of rock mass on anchorage effect of high slope under strong unloading condition (41472262)", and thanks for that.

The first chapter of this book is mainly written by Profs. Zhou Jianting (jtzhou@cqjtu.edu.cn) and Ren Qingyang (qyren@cqjtu.edu.cn), and the rest chapters are written by Prof. Ren Qingyang.

Chongqing, China

Qingyang Ren
Jianting Zhou

Brief Introduction of the Book

The constitutive model of soil is the basic key problem in the theory and practice of geotechnical engineering. This book abandons the traditional idea of looking for plastic potential as the only way to model, and adopts numerical method to model. Firstly, the triaxial compression tests of expansive soil, sand and clay under different stress paths are introduced; then the elastoplastic constitutive equations of expansive soil, sand and clay under various stress paths are established by numerical modeling method; finally, the constitutive equations are embedded in the finite element program, and verified by comparing the finite element calculation results of the triaxial test soil samples with the corresponding test results, so the accuracy of the model is obtained.

This book can be used as a reference for teachers, scientific researchers and engineering designers of colleges, universities and scientific research institutes engaged in geotechnical engineering and engineering geology.

Contents

Symbols

σ	Stress
ε	Strain
E	Modulus of elasticity
G	Shear modulus
K	Bulk modulus
τ	Shear stress
P	Average normal stress
q	Generalized shear stress
σ_1	Maximum principal stress
σ_2	Intermediate principal stress
σ_3	Minimum principal stress
υ	Poisson's ratio
ε_{e}	Elastic strain
ε_{p}	Plastic strain
ε_{υ}	Bulk strain
$\bar{\varepsilon}$	Shear strain
$\varepsilon_{\upsilon}^{\mathrm{p}}$	Plastic body strains
$\varepsilon_{\upsilon}^{\mathrm{e}}$	Elastomer strain
$\varepsilon_{s}^{\mathrm{p}}$	Plastic shear strain
u	Pore pressure
w	Water content
ρ	Density
e	Void ratio
OCR	Overconsolidation ratio
c	Cohesion
φ	Internal friction angle
$W1$	lLiquid limit
$W\mathrm{p}$	Plastic limit
$I\mathrm{p}$	Plasticity index
f	Yield function
F	Force
D	Deformation

DCTC Consolidated drained shear triaxial compression test
DPTC Consolidated drained shear triaxial compression test
UCTC Consolidated undrained shear triaxial compression test
DRTC Consolidation drainage shear and confining pressure reduction triaxial compression test
ERTC The ratio of main stress to consolidated drainage triaxial compression t

Chapter 1
Introduction

Constitutive relationship is a mathematical model reflecting the macroscopic properties of matter. The most familiar constitutive relations reflecting pure mechanical properties are Hooke's law, Newton's law of internal friction, Saint Venant's law of ideal plasticity, etc., and those reflecting thermodynamic properties are Clapeyron's equation of state of ideal gas, Fourier's equation of heat conduction, etc. The constitutive equation is the mathematical expression of constitutive relation. In many literatures, constitutive relation and constitutive equation are often not distinguished. The establishment of constitutive equation is one of the important contents in the study of rational mechanics.

1.1 Constitutive Relationship of Soil

The constitutive relation of soil is the relation between stress tensor and strain tensor. In general, it refers to a group of relations that relate the deformation parameters of soil to the parameters of internal force. Specifically speaking, it refers to a set of relations, also known as constitutive equation, which links the strain tensor of deformation with the stress tensor. For different soil, there are different constitutive relations under different deformation conditions, also known as different constitutive models, which is a comprehensive reflection of the macro mechanical properties of soil.

Since 1963 when Roscoe and others proposed the Cam clay model, the research of soil constitutive relationship has developed rapidly. Especially in the past 20 years, the numerical method to solve engineering problems has been basically mature and widely used, which makes the complex engineering problems can be solved practically. However, in solving these engineering problems, the key problem is still the engineering characteristics of soil, that is, the constitutive relationship between stress and strain of soil and time, that is to say, the constitutive relationship of soil is also the

Q. Ren and J. Zhou, *Numerical Modeling of Soil Constitutive Relationship*,
https://doi.org/10.1007/978-981-16-3231-0_1

primary problem in the numerical analysis of soil mechanics and geotechnical engineering. The purpose of this book is to put forward the numerical modeling method of soil constitutive relation based on the discussion of soil constitutive relation of stress, strain and time.

1.2 Mechanical Properties of Soil

Soil is a kind of sediment formed in various natural environments by continuous and solid rock particles with different sizes formed under weathering and different transportation methods. In the long geological age, many types of rocks and soils have been formed due to various internal and external forces. The rock undergoes weathering, denudation, transportation and deposition to form soil, while the soil undergoes compaction and consolidation, and cementation and hardening can also generate rock again.

The material components of soil include solid mineral particles as soil skeleton, water in pores, dissolved substances and gas. Therefore, soil is a three-phase system composed of particles (solid phase), water (liquid phase) and gas (gas phase) (Fig. 1.1). The particle size and mineral composition of all kinds of soil are very different, and the quantity proportion of the three phases of soil is not the same, and the complex physical and chemical effects of soil particles and the surrounding water occur. Therefore, in order to study the properties of soil, it is necessary to understand the three-phase composition of soil and the structure and structure of soil in natural state.

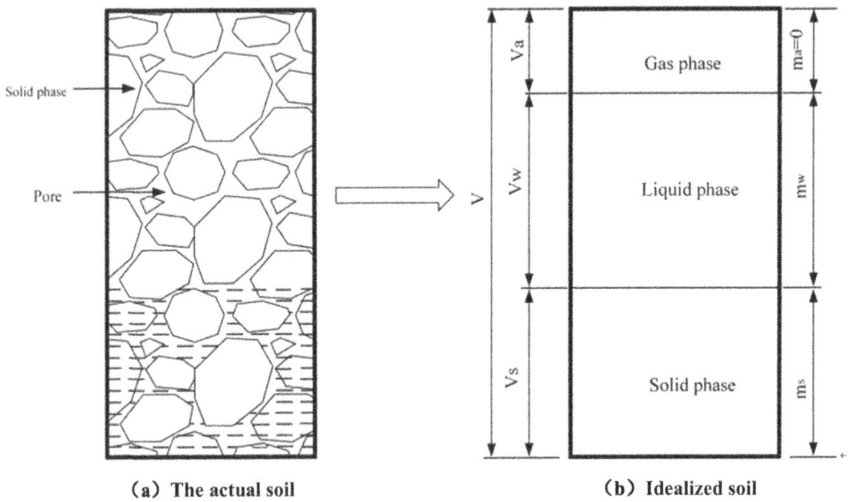

Fig. 1.1 Three phase diagram of soil

According to whether there is cohesive force between soil particles, soil can be divided into cohesive soil and non cohesive soil. According to the stress history, clayey soil can be divided into normal consolidated clay and overconsolidated clay. The overconsolidated ratio of normal consolidated clay is OCR = 1, and the overconsolidated clay OCR > 1. Cohesionless soil mainly includes gravelly soil, sandy soil, silt, etc.

The track of stress state change in the process of soil loading is called stress path, and the track of strain state change is called strain path. The whole historical process of stress change in soil from the beginning of formation to a certain point when it is studied is called stress history. The pore ratio, water content, structure and material composition, stress history and stress path all affect the mechanical properties of soil. It mainly includes the following features.

1. **Nonlinearity of stress-strain relationship**

The macro deformation of soil is not due to the deformation of particles themselves, but to the change of the position between particles. In this way, under different stress levels, the strain increment caused by the same stress increment will not be the same, that is to say, it shows strong nonlinearity.

Figure 1.2 shows the general results of the conventional triaxial compression test of soil. Among them, the solid line represents dense sand or overconsolidated clay, and the dotted line represents loose sand or normally consolidated clay. It can be seen from the figure that the stress of normal consolidated clay and loose sand increases with the increase of strain, but the rate of increase becomes slower and slower, and finally approaches an asymptote; in the test curve of dense sand and over consolidated soil, the stress begins to increase with the increase of strain, and after reaching a peak, the stress decreases with the increase of strain, and finally tends to be stable. In plastic theory, the former is called strain hardening (or work hardening), and the latter is called strain softening (or work softening). The strain softening process is actually an unstable process, sometimes accompanied by the appearance of localized shear band of stress, and its stress–strain curve is sensitive to some influencing factors. Because

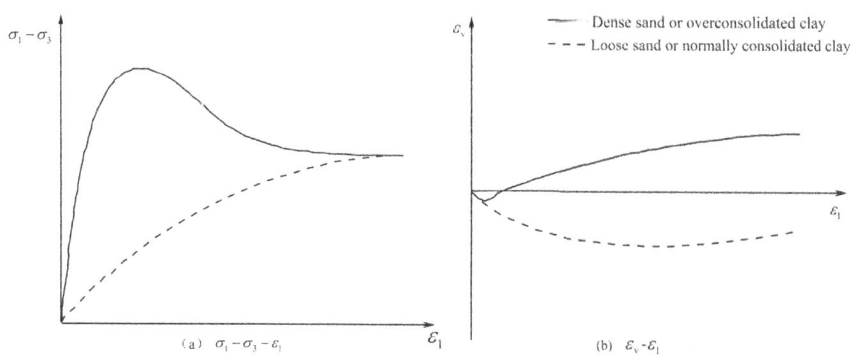

Fig. 1.2 Typical triaxial compression test curve of soil with different compactness

the stress–strain relationship is not a single value function, the mathematical model to reflect the strain softening of soil is generally complex and difficult to describe accurately; the numerical calculation method to reflect the strain softening is also difficult.

2. **Dilatancy of soil**

When the soil is compressed in the same direction or in the same ratio, the porosity is reduced, so the larger volume compression occurs. Most of this volume compression is irreversible. It can be seen from Fig. 1.2b that in triaxial test, the increase of deviating stress of dense sand or overconsolidated clay results in the increase of axial strain, but in addition to a small amount of volume compression (positive strain) at the beginning, obvious volume expansion (negative strain) occurs. In the conventional triaxial compression test, the average increment of principal stress Δ P = / 3 is always positive in the loading process, which can not be the elastic rebound of volume. Therefore, the volume strain can only be caused by shear stress, which is called shear dilatancy. The generalized dilatancy refers to the volume change caused by shear, including volume expansion and volume contraction. The latter is often referred to as shearing. In fact, the dilatancy of soil is caused by the change of the mutual position of soil particles caused by the shear stress, which makes the arrangement change and the pores between particles increase (or decrease), so that the volume changes.

3. **Elastoplasticity**

When the soil is loaded and unloaded to the original stress state, it will not return to the original strain state. Part of the strain is recoverable, part of the strain is non recoverable plastic strain, and the latter often accounts for a large proportion, which can be expressed as. Where, represents elastic strain, represents plastic strain. As shown in Fig. 1.3, the monotonic loading test curve is represented by dotted line; the cyclic loading test curve is represented by solid line. It can be seen that every stress cycle has recoverable elastic strain and irrecoverable plastic strain, namely permanent deformation.

For undisturbed soil with strong structure, such as hard clay, its deformation may be almost "elastic" within a certain stress range, and plastic deformation will occur only when it reaches a certain stress level, i.e. the yield condition. Generally, the elastic and plastic deformation of soil occur almost at the same time in the process of loading, and there is no obvious yield point, so it is also called elastic–plastic material.

Another characteristic of soil in the stress cycle is the existence of hysteresis loop. In Fig. 1.3, the stress–strain curve drops sharply at the initial stage of unloading. When it reduces to a certain deviation stress, the unloading curve slows down, and then the loading curve begins to steep and then slows down. This forms a hysteresis loop,

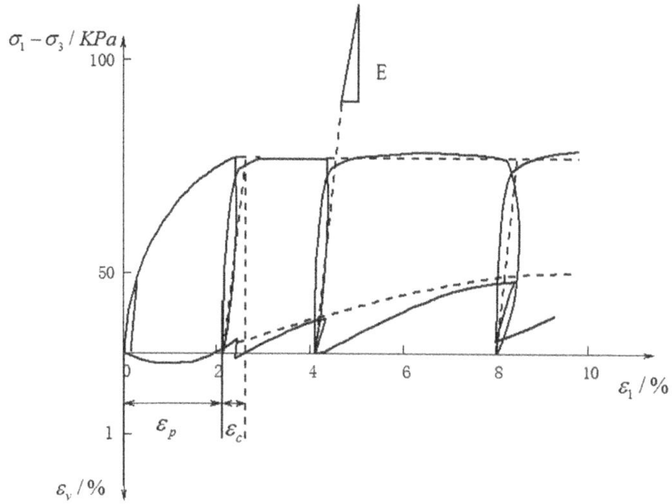

Fig. 1.3 Triaxial compression test curve under monotonic and cyclic loading

and the closer the failure stress is, the more obvious the phenomenon is. Another noteworthy phenomenon in Fig. 1.3 is the volume shrinkage of the specimen during unloading. Because the average principal stress P of soil is reduced during unloading, it is obvious that the unloading volume shrinkage can not be explained by elastic theory. It is believed that this is mainly due to the recoverability of soil shear expansion deformation and the change of soil structure caused by loading. In a word, the deformation of soil is not completely elastic even in the unloading reloading process under the same stress path.

4. **Anisotropy and soil structure**

The so-called anisotropy refers to the different physical and mechanical properties of materials in different directions. In the process of soil deposition, the needle, sheet and rod shaped particles with length width ratio greater than 1 tend to be arranged horizontally under the action of gravity and are in a stable state; in addition, in the subsequent consolidation process, the vertical stress caused by the vertical gravity of the overlying soil and the horizontal stress generated by the horizontal earth pressure are not the same, and the anisotropic consolidation will also produce the soil. The anisotropy of soil is mainly manifested as transverse isotropy, that is to say, the properties in all directions of the horizontal plane are basically the same, while the vertical and horizontal properties are different. The anisotropy of soil can be divided into initial anisotropy and induced anisotropy. The anisotropy caused by natural deposition and consolidation can be classified as initial anisotropy. In the indoor gravity field, various sample preparation processes will also make the soil sample have different degrees of initial anisotropy, as shown in Fig. 1.4.

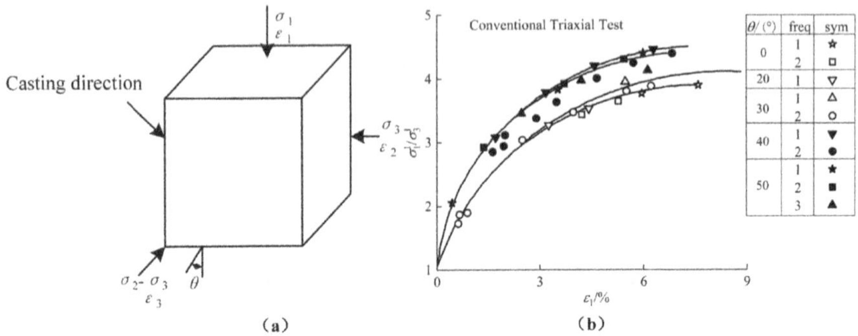

Fig. 1.4 Anisotropy of sand

5. Rheology

The stress–strain relationship of cohesive soil is affected by time, not only the dissipation of pore pressure and the consolidation of soil based on the principle of effective stress, but also the rheology of soil. The phenomena related to the rheology of soil are creep and stress relaxation of soil (Fig. 1.5). Creep refers to the phenomenon that the strain increases gradually with time under the condition that the stress state remains unchanged; stress relaxation refers to the phenomenon that the stress in the material decreases gradually with time while the strain remains unchanged. Under the action of a certain constant stress, the strain of soil increases continuously, but when the stress value is small, the deformation of sample tends to be stable gradually; when the constant stress is large, the strain will accelerate suddenly after the relative stability, and finally reach creep failure. This creep strength is lower than that of conventional test, sometimes only about 50% of the latter. The creep of clayey soil increases with the increase of plasticity, activity and water content.

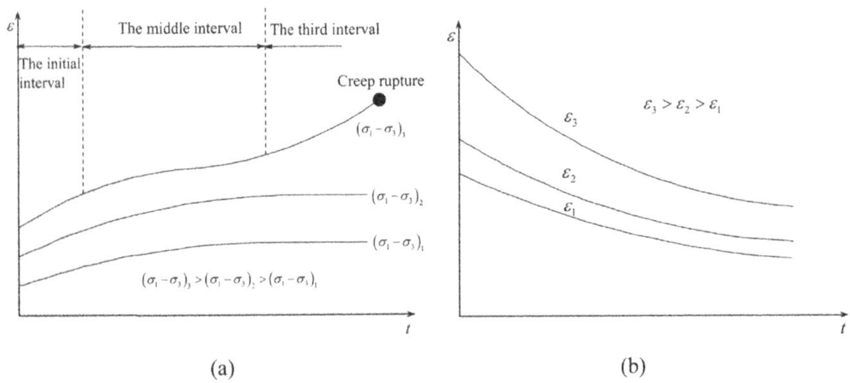

Fig. 1.5 Creep and stress relaxation of soil

Under the condition of confining compression, the compression due to the rheology of soil is called secondary consolidation, long-term secondary consolidation.

The normal consolidated soil shows the characteristics of overconsolidated soil, which is called as overconsolidated soil or "old clay". The compression curve of structural soil is shown in Fig. 1.6.

6. Influence of stress level, stress history and stress path

(1) Effect of stress level

The so-called stress level generally has two meanings: one is the absolute value of the confining pressure; the other is the ratio of the stress (usually shear stress) to the failure value, that is s = q/QF. The stress level here refers to the confining pressure.

As shown in Fig. 1.7, with the increase of σ 3, the strength and rigidity of sand are obviously improved, and the shape of stress–strain curve is also changed. Under very high confining pressure, even if the soil is very dense, it is similar to the stress–strain curve of loose sand: there is no dilatancy and strain softening phenomenon. It should be pointed out that the shear strength of excavated soil increases with normal stress or confining pressure, but the stress ratio or internal friction angle of sand at the time of failure often decreases with the increase of confining pressure. The phenomenon that the deformation modulus of the soil increases with the confining pressure is also known as the compressibility of the soil. Because soil is composed of broken particles, the restraint provided by confining pressure is very important for its strength and rigidity. This is also one of the important characteristics of soil different from other materials.

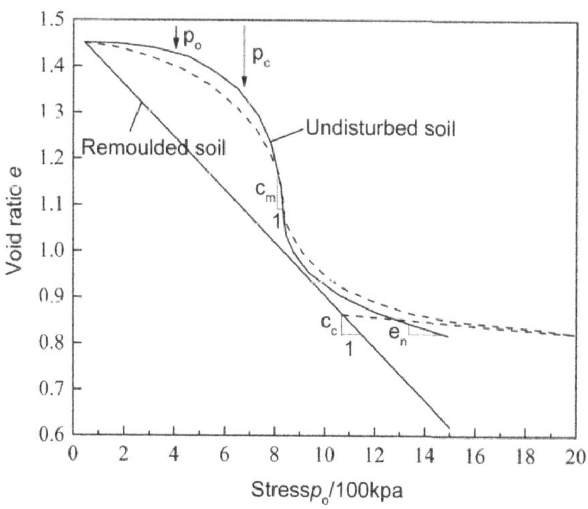

Fig. 1.6 Compression curve of structural soil

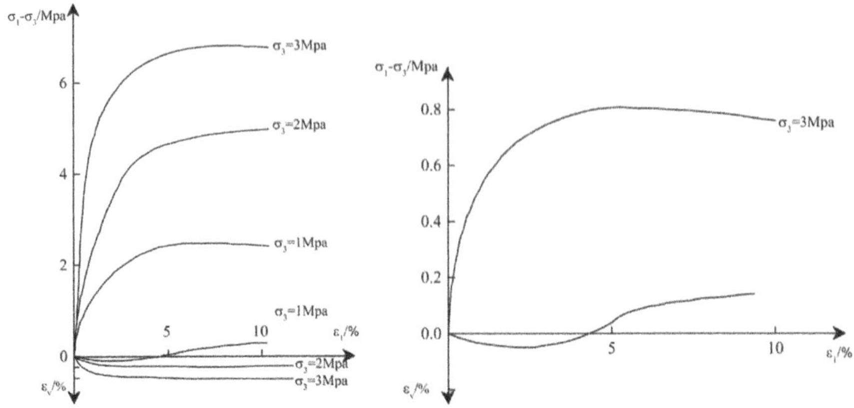

Fig. 1.7 Triaxial compression test curve of medium dense sand under different confining pressures

(2) The influence of stress history

Stress history includes not only the consolidation and crustal movement of natural soil in the past geological years, but also the stress process of soil in laboratory (or in engineering construction and operation). For cohesive soil, it generally refers to its consolidation history. If the maximum preconsolidation pressure (effective stress) of cohesive soil in its history is greater than the consolidation pressure at present, then it is overconsolidated clay. If the current consolidation pressure is the maximum consolidation pressure in its history, then it is normally consolidated soil. As mentioned above, the rheology of the soil makes the cohesive soil under long-term load. Although the consolidation stress has not changed in history, the secondary consolidation makes the soil show the character of over consolidation. This is also an effect of stress history.

(3) Influence of stress path

It can be seen from the stress–strain curves of triaxial tests with different stress paths that they are obviously affected by stress paths. For example, a triaxial test of two stress paths for loose sand is shown in Fig. 1.8. Their starting point a and ending point B are the same, but path 1 is from a-1-b; path 2 is a-2-b. A large axial strain occurred in path 1. This is because the stress at point 1 is closer to the failure line than that at point B, which results in larger axial strain. Wood used box true triaxial apparatus to conduct shear test on the remolded saturated clay under the condition of isotropic consolidation and OK direction, and then continued the test on the different stress paths of KM, kn and KL starting from point K to obtain the strain path. It can be seen that the direction of strain path is consistent with the direction of stress path when the load is in the original direction of OK. However, when the stress path turns, the cohesive soil seems to have "memory" for the path just passed, or the strain path has inertia along the OK direction. Only after a long distance on the new stress path, the direction of the strain path gradually approaches.

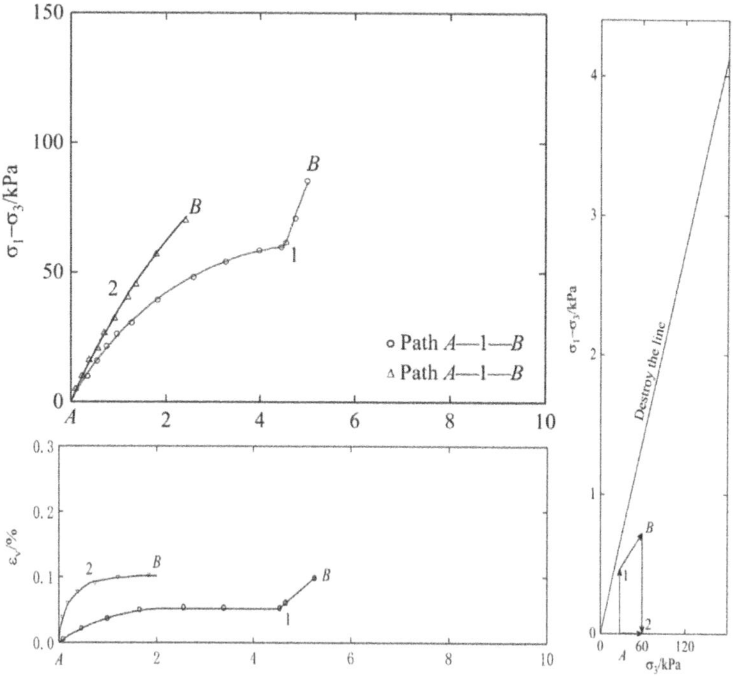

Fig. 1.8 Stress–strain curve of loose sand under different stress paths

1.3 Development and Current Situation of Soil Constitutive Model

The properties of soil can be expressed by several kinds of constitutive models, such as various modulus, elastic formula, super elastic formula, sub elastic formula, internal time theory and plastic formula, among which plastic based model is the most common. In addition to the basic physical properties of soil, such as void ratio, water content, soil structure and minerals, the current and prior stress state and stress path of soil also have an important impact on the properties of soil, sometimes the impact of stress path is even decisive.

The purpose of the constitutive model is to get the incremental relationship of stress–strain, and the parameters can only be determined in the main space, which requires the application of certain assumptions, and different assumptions constitute different constitutive theories. The methods of building soil constitutive model can be divided into analytical method and numerical method. In recent 20 years, the research of analytical method mainly focuses on the constitutive model based on modified Cambridge model, the constitutive model based on generalized potential theory, the constitutive model considering internal structure and the constitutive model considering the coupling effect of temperature water soil force. The research of numerical method mainly focuses on damage theory model, neural network model and other mathematical statistics based models.

In the following, the research of geotechnical constitutive relationship at home and abroad in recent years is summarized.

1.3.1 Analytical Method

1. **Constitutive model based on modified Cambridge model**

Roscoe (1968) [1] of the University of Cambridge, UK, established a Cambridge model to describe the normally consolidated soil. The model uses the cap yield surface and related flow criteria, and takes the plastic soil strain as the hardening parameter. Roscoe and Burland modified the Cambridge model, proposed the modified Cambridge model, and obtained the elliptical yield track. Mita et al. (2004) [2] proposed a new soil Hvorslev MCC constitutive model. Compared with the traditional queuing critical model (MCC), this model overcomes the disadvantage that MCC is not suitable for three-dimensional space, and can better predict the peak deformation of soil at the supercritical point. Woodward and Berenji (2001) [3] extended Sekiquchi OHTN model to clay, silt and sand in three-dimensional space by adding a variable stress tensor and hardening parameter H. The model can reflect the phenomenon of dilatation and shrinkage of soil caused by initial stress. Oettl et al. (1998) [4] simulated the stress state of tunnel excavation and concrete lining on the basis of plane finite element analysis. Through the application of four soil constitutive models - linear elastic model, elastic–plastic model, Mohr Coulomb criterion and cap model in the simulation, the different effects of the four models on the prediction of tunnel displacement and stress distribution are compared. Lade and Inel (1997) [5] believe that there is a kinematic hardening constitutive model in the stress reversal test of sand. In this model, the interaction between rotation and yield surface is considered to fit the experimental data, and the behavior from isotropy to kinematic hardening under monotonic loading and tension is well simulated.

Yao et al. (2009, 2011a, 2015) [6–8] proposed a unified hardening model of soil based on the modified Cambridge model, with the transformation stress method and unified hardening parameters as the basic elements. In this model, the transformation stress method based on SMP criterion, lade criterion or generalized nonlinear strength theory is used to realize the organic combination of the model and strength criterion and the three-dimensional of the model. Based on the concept of true strength and the determination method of potential strength, the theoretical calculation formula of the critical state stress ratio of unsaturated clay at different temperatures is derived, and the normal considering the temperature effect is established The constitutive model of consolidated unsaturated soil is extended to over consolidated unsaturated soil, and the generalized nonlinear strength criterion and the transformation stress three-dimensional method satisfying the thermodynamics law are proposed, thus realizing the reasonable three-dimensional of unified hardening (Uh) constitutive model. Chen et al. (2015) [9] introduced the overconsolidated remolded soil constitutive model into the construction of the elastic–plastic constitutive model of marine sedimentary soft clay, taking into account the characteristics of the tensile strength and evolution of marine sedimentary undisturbed soft clay, the strength envelope of soft clay, and further modified the expression to make the model more consistent with the strength and deformation characteristics of marine undisturbed soft clay. Yao et al. (2011a) [10] combined the Barcelona constitutive model with the uh constitutive model of overconsolidated soil proposed by Yao Yangping et al. To make it suitable for overconsolidated unsaturated soil. The model can reflect the hardening, softening, shear shrinkage, shear expansion characteristics of overconsolidated unsaturated soil and the influence of different stress paths on the deformation characteristics of overconsolidated unsaturated soil. Based on the critical state theory, Yao Yangping and Yu Yani (2011) introduced state parameters and parameters of modified yield function to adjust the hardening parameters and shear expansion equation, and used the power function relationship between sand isotropic consolidation line and critical state line in void ratio and effective stress to describe the stress–strain response of sand in a large density and effective stress range. Yao Yangping and Yu Yani (2011) [11] based on the UH model of overconsolidated soil, extended the asymptotic state constitutive model applicable to saturated sand to the overconsolidated soil constitutive model considering the asymptotic state characteristics, and realized the three-dimensional model by using the transformation stress method. Hou Wei et al. (2008) [12] put forward a unified hardening overconsolidation model which can describe the shear shrinkage, shear expansion, hardening and softening characteristics of overconsolidated soil on the basis of the lower loading surface and the theoretical framework of Cambridge model, and introduced the parameters related to the soil structure into the overconsolidated soil model to establish the structural overconsolidated soil model.

The Cambridge model creates the critical state theory of soil mechanics, but it is based on the construction of a single yield surface, which can only reflect the shear shrinkage, but not the shear expansion. Therefore, some yield surfaces, double yield surfaces and triple yield surfaces are proposed to reflect the characteristics of soil such as shear shrinkage and shear expansion, hardening and softening, and critical state.

2. **Constitutive model based on generalized potential theory**

Yang Guanghua [13] and others began to explore new theoretical methods of soil constitutive model since 1988. They established the theoretical basis of the constitutive model from the mathematical point of view, established a unified mathematical connection with the existing modeling theory, and proposed the generalized potential theory.

Based on the generalized potential theory, Wen Yong [14] proposed solutions to the limitations of the traditional plastic potential theory in describing the non uniqueness of the increment direction of plastic strain and the non coaxial problem of soil, and proposed solutions to the limitations of Duncan Chang model based on the generalized Hooke law and the hyperbolic function fitting test curve. Tang Shuai (2013) [15] studied the decomposition criterion of plastic strain increment, deduced the expression method of stress–strain relationship from the mathematical basis, abandoned three basic assumptions of yield criterion, flow rule and hardening rule in the process of building the model, and established a multiple potential surface model. Yao Jie (2010) [16] analyzed the characteristics of the constitutive matrix of each model family based on the generalized potential theory, studied the decomposition criteria of plastic strain increment, proposed and established the constitutive model considering quasi elastic plastic deformation, and decomposed the traditional non recoverable plastic strain increment into the quasi elastic plastic deformation part with elastic strain characteristics and the pure plastic part conforming to the traditional plastic theory. Zhou Aizhao and Lu Tinghao (2008) [17], based on the generalized potential theory, regard the contact surface problem of soil and structure as a two-dimensional problem in space, use two linearly independent potential function gradient vectors to represent the vector composed of two components of plastic strain increment, and use the plastic state equation to replace the traditional yield surface, and establish a new elastoplastic model of the contact surface of double potential surface.

According to the generalized potential theory, the modeling ability of geotechnical materials can be extended. The generalized potential theory has the advantages of clear mathematical principle and convenient modeling. It does not need to be based on the plastic postulate, but also contains the traditional classical theory as its special case. It should be a more promising method for further development and improvement.

3. **Constitutive model considering internal structure**

For the anisotropic consolidated clay, the influence of the internal structure of the material on the constitutive relationship is very obvious. The traditional isotropic consolidated soil model can not truly and accurately reflect the response of the material. It is necessary to introduce the rotation hardening theory to reflect the anisotropy of the material on the yield surface to a certain extent.

Yang et al. (2015) [18] in the constitutive relation of rotational hardening, the non related flow relation of materials is realized by explicitly defined plastic potential function. The evolution rule of rotational hardening is controlled by plastic body strain and plastic shear strain together. The initial rotation angle is no longer equal to the initial stress ratio. A new set of formulas is proposed to determine the initial rotation angle of yield surface and plastic potential surface. Sang and Salgado (2015) [19] proposed a sand boundary interface model considering fabric and its evolution, and proposed a new phase transition line (PTL) in e-p space to reflect shear expansion and shear shrinkage, which is determined by fabric. Dafalias and Taiebat (2013) [20] introduced rotational hardening on the basis of modifying the Cambridge model. By explicitly defining the dissipation potential function to describe the non related flow relationship in the loading process, the value range and critical state of the rotational hardening formula were discussed. The initial rotation angle was corrected by K0 test. It was pointed out that the rotation degree of the yield surface reflected the anisotropic level of the material According to the theory of anisotropic critical state, materials should have unique critical state fabric, so the degree of rotational hardening must meet the unique requirements. Two kinds of rotation hardening laws are put forward: one is that the yield surface in critical state returns to horizontal symmetry; the other is that the yield surface has critical state rotation angle. Sun et al. (2004) [21] proposed a dynamic model to describe the deformation of fine clay under the action of pile foundation. It is also pointed out that the axial bearing capacity NY is related to the depth of the foundation. Manzari and Dafalias (1997) [22] put forward the double-sided plastic expression of soil under the frame of critical state soil mechanics, established the constitutive model of sand in the full stress space, and applied the model to simulate the deformation of sand under monotonic and cyclic loading, drainage and undrained conditions.

Chen Yanni (2016) [23] proposed a new dissipative potential function considering the shape change and rotation of the yield surface, and obtained the yield function of the real stress space based on the analysis of the dissipative potential function of the modified Cambridge model and the α non correlation model. At the same time, the boundary interface theory was introduced to realize the elastic–plastic response of the overconsolidated soil at the initial stage of loading. Starting from Wheeler's s s-clay1 model, Wang Lizhong and Shen Kailun (2008) [24] studied the comprehensive effect of plasticity anisotropy, structure and evolution rule of clay, introduced the concept of rotation limit curve, and added the parameter describing the evolution rate of anisotropy to explore the influence of rotation hardening on the mechanical deformation of soil. Kong Liang et al. (2008) [25] improved the modified Cambridge model on the basis of rotation hardening and unified hardening theory. The model only added one soil parameter compared with the modified Cambridge model, but it can reflect the soil deformation characteristics of monotonic loading and cyclic loading; it is applicable to both clay and sand. Wei Xing and Huang Maosong (2007) [26] introduced rotation hardening parameters and shape parameters on the basis of modifying Cambridge model, improved Wheeler's rotation hardening evolution formula, and ensured that rotation hardening increment in critical state was 0. Kaolin and Boston clay were simulated respectively. Jie Yuxin et al. (2004) [27] proposed a plastic constitutive model of anisotropic soil with non associated flow rules, which successfully simulated the mechanical response of normally consolidated clay under undrained conditions; Wang Jianhua and Yaominglun (1996) [28] established an incremental elastoplastic model in the form of total stress according to the theory of non isotropic hardening modulus field for the dynamic deformation characteristics of soft clay The method of determining model parameters by dynamic triaxial test results is presented. Huang Wenxi et al. (1981) [29] proposed that the work hardening law and yield function can be directly determined from the test data, so as to meet the orthogonality and ensure the uniqueness of the solution. Xu Riqing et al. (1996) [30] used the smooth closed egg function as the yield surface and the edge interface, and established the internal relationship between them with the strength development degree, so as to establish the edge interface constitutive model.

At present, there is no unified system for the study of rotational hardening at home and abroad. Due to the lack of experimental and theoretical basis, the mathematical expression of the law of rotational hardening is quite different, which can not be unanimously recognized, and needs further development.

4. **Constitutive model considering coupling effect**

The soil in natural state is under certain temperature, hydraulic condition and complex stress condition, but different temperature and hydraulic condition have great influence on the stress–strain state of soil, so it is unscientific to study the constitutive model of remolded soil in laboratory.

Oka et al. (2002) [31] used the elastic viscoplastic constitutive model to study the influence of microstructure of saturated clay immersed in water on its dilatancy and permeability. Thomas and Cleall (1999) [32] put forward a constitutive model which can reflect the heat flow, humidity and gas deformation in the expansive soil, and carried out the heat simulation experiments of humidity, water pressure and mechanics related to the expansive soil, which proved that the model can simulate the plastic deformation of the expansive soil in the dry wet cycle state. Rosa and Wulfsohn (1999) [33] proposed a plastic constitutive model to describe the dynamic response of narrow tools to cultivated land. According to Stallebrass and Taylor (1997) [34], the stress–strain response of overconsolidated soil depends not only on the current and past stress history, but also on the current and past stress paths. An elastic–plastic constitutive model is proposed to predict the deformation of overconsolidated soil. Katti and Desai (1996) [35] proposed a constitutive model of cohesive soil under undrained cyclic loading. The model considers the influence of dynamic water pressure, and can truly predict the strain of sand under initial stress or disturbed state.

Kong Lingming and Yao Yangping (2015) [36] introduced the influence of temperature change on clay volume and strength parameters into the isotropic stress–strain-time relationship of overconsolidated soil, established the stress–strain-time–temperature relationship under the condition of isotropic stress, deduced the hardening law of yield surface, and combined it with the yield equation and flow law of the unified hardening model of overconsolidated soil, established the heat of overconsolidated soil Viscoelastic plastic constitutive model. Ma Tiantian et al. (2014) [37] established the coupling constitutive model of unsaturated soil on the basis of modifying the Cambridge model by using the generalized Mises criterion, applied the Songgang Zhongjing (SMP) criterion to the model to make it reasonable three-dimensional, and effectively extended the model from the axisymmetric stress state to the general stress state. Liu Yan et al. (2014) [38] established the numerical model of the unsaturated soil coupling model by using the implicit integration method, obtained the consistent tangent modulus of the hydraulic mechanical coupling unsaturated soil, considered the influence of saturation in the hardening equation, and considered the influence of plastic body change in the soil water characteristic curve, so that the model can reflect the capillarity and elastoplasticity in the unsaturated soil Coupling behavior of deformation phenomenon. Ma Tiantian et al. (2014) modified the coupling constitutive model of capillary hysteresis and plastic deformation of unsaturated soil to allow for the influence of residual air content. Jin Xu and Zhao Chenggang (2011) [39] proposed the concept of coupling disturbance variable reflecting the influence of external force and infiltration, established the constitutive model applicable to unsaturated structural soil, considered the coupling effect of water hysteresis and soil deformation, and the influence of saturation change on structural deterioration, and gave the evolution equation of coupling disturbance variable. The first mock exam of Wang Na (2010) [40] is based on the generalized effective stress principle of unsaturated soils and the volume change model of unsaturated soils. A elasto-plastic constitutive model which can simulate the hydraulic and mechanical coupling of unsaturated soils is proposed. Yao Yangping and Zhu Enyang (2010) [41] constructed a coupling stress which can comprehensively consider the interaction of the average stress and the generalized shear stress based on the interaction of the average stress and the generalized shear stress of the soil when they cause the plastic strain. Based on the analysis of the existing stress–strain test rules of the clay, the one-dimensional constitutive relationship between the plastic volume strain and the coupling stress of the clay was deduced Department.

At present, the effects of coupling on constitutive model are mainly hydraulic soil coupling, temperature stress coupling and stress–strain coupling, which are mainly the products in some specific environment. They are not universally applicable and have certain limitations. However, the research on multi field coupling of temperature, water stress, stress and strain has started, and it is expected to establish multi field coupling cooperation Using the unified theoretical basis of soil constitutive model.

5. **Constitutive model established by other analytical methods**

Shively (2015) [42] proposed a critical state equation considering the effect of initial void ratio, and established a constitutive model of coarse-grained soil based on state parameters by constructing plastic modulus. Sheng et al. (2008) [43] put forward SFG model, which has simple parameters, wide application and can describe the influence of net stress and suction interaction on soil body change, but the model can not describe the plastic expansion phenomenon of expansive soil. Dafalias and Manzari (2004) [44] established a sand plastic model with uniaxial stress control consistent with critical state, and extended it from uniaxial state to multiaxial state. Calvello et al. (2004) [45] studied how to select appropriate parameters and optimize the accuracy of soil constitutive model with back analysis technology. Taking the Chicago Quaternary ice age clay elastoplastic model as the research object, discussed the factors affecting the accuracy of the model. Ensan et al. (2003) [46] established a macro elastic–plastic constitutive model of multi-layer soil, which considered the stress relaxation of soil interface, and showed the stress–strain state of soil interface through numerical simulation. Li and Ding (2002) [47] proposed a three-dimensional microstructure deformation model under cyclic loading of sand on the basis of studying the three-dimensional micro mechanical properties of fine-grained soil. Dafalias et al. (2002) [48] studied the triaxial test of textile reinforced soil under small strain cyclic load, and proposed the nonlinear elastic constitutive model of soil. Zhang et al. (2001), Wan and Guo (1998), Boulon and Alachaher (1995) [49–51] introduced several new constitutive models of sand. Oettl et al. (1998), Chowdhury et al. (1999), Stallebrass and Taylor (1997), Wheeler and Sivakumar (1995) [4, 34, 52, 53] introduced several new constitutive models of clay. Goto et al. (1999) [54] proposed to describe a new nonlinear model with three material parameters. All parameters can be obtained from the test data. The calculated results are basically consistent with the stress–strain relationship measured by the test. Ortiz and Pandolfi (1999) [55] proposed a finite deformation constitutive model of cohesionless granular soil. Li et al, Adachi et al. (1998) [56, 57] put forward the elastic–plastic constitutive model of geological materials under strain softening. The model not only considers the stress history, but also considers the influence of time. Through the test results of the model on soft sedimentary rock, it shows that the model can not only describe the influence of time, but also consider the strain softening behavior of geological materials. Krogsboll (1998) [58] put forward a soil constitutive model, which can describe the properties of soil under different deformation rates, and give the influence of time effect on the constitutive relationship of cohesive soil. Fox and Berles (1997) [59] put forward a piecewise linear finite difference model CS2 model under large strain of consolidated soil. Niemunis and Krieg (1996) [60] established the viscoplastic constitutive model of normally consolidated soil under uniaxial compression, and improved the one-dimensional creep theory. Bardet (1996) [61] used the concept of point memory to describe the stress–strain curve of soil under monotonic load, which was successfully applied to simulate the strain state of several sand and clay under cyclic loading. Leidwanger-Rabis et al. (1995) [62] proposed an elastic viscoplastic model to predict the long-term settlement of mega structures, and modified the model parameters that affect the settlement sensitivity. Duncan and Chang (1970) [63] combined Duncan Chang model with Drucker Prager/Mohr Coulomb yield criterion elastic–plastic constitutive model, and put forward a nonlinear elastic–plastic model, which overcomes the shortcomings of traditional elastic–plastic model that can not consider the non-linear behavior of geotechnical materials before plastic yield and is not applicable to the stress level close to yield or failure state.

Based on the triaxial test of coarse-grained soil, Wei Guomin et al. (2016) [64] established a dual yield surface model of coarse-grained soil based on state parameters, and deduced its stress–strain expression. The model can reflect the peak strength, dilatancy, strain hardening or softening laws of coarse-grained soil under different initial void ratio conditions. Zhang Bo et al. (2015) [65] based on the generalized plastic theory and the concept of hvorslev surface and overconsolidation parameters, established a generalized plastic constitutive model which can reflect the deformation characteristics of overconsolidated soil. Li Wugang et al. (2015) [66] deduced the neutral loading surface (NL yield surface) equation of expansive soil under the framework of SFG model, established the constitutive model of expansive soil, which reflected the micro level deformation of soil with the total deformation, no longer distinguished the structural deformation of micro level and macro level, and reduced the parameters of the constitutive model of unsaturated expansive soil. Based on the existing unsaturated soil model framework, Liu Yan et al. (2013) [67] introduced the influence of gas phase dissipation, considered the influence of the change of closed gas pressure in the hardening equation, and established an elastic–plastic constitutive model for unsaturated soil under the condition of high saturation using the generalized effective stress principle. Based on the SFG model, Liu Min (2011) [68] adopted the effective stress derived from the expression of work as the stress state variable, selected the continuous model which can reflect the relationship between the current void ratio and saturation, established the modified SFG model, carried out simulation calculation through MATLAB, and verified the applicability of the model. Shen Cunke et al. (2010) introduced plastic work into the energy equation of the process of soil stress and deformation, deduced the soil flow rule; adopted the straight-line yield track and non associated flow rule, calculated the hardening function by the undrained stress path, and established a coarse-grained soil constitutive model considering particle breakage. Jia Yufeng et al. (2010) [69, 70] established the stress–strain relationship considering the energy consumption of particle crushing based on the triaxial test, and deduced the unified constitutive model considering the shear dilatancy of coarse-grained soil with particle crushing by using the correlation flow law. Sun Haizhong and Huang Maosong (2009) [71] improved the traditional constitutive model of fine-grained soil, and established an elastic–plastic constitutive model which can better describe the strain softening characteristics and shear dilatancy of coarse-grained soil by adopting the form of double yield surface. Li Guangxin et al. (2008) [72] introduced moisture content into the hardening parameters of Tsinghua model to establish the Tsinghua elastic–plastic model of unsaturated soil, studied the relationship between stress–strain and strength of soil under the condition of unsaturated humidification, and verified that the model can predict the relationship between stress-deformation and strength of unsaturated soil with different moisture content by the method of adding ice chips in dry soil. Miao Linchang (2007) [73] used the concept of average soil skeleton stress to deduce the relationship between the stiffness parameters of unsaturated soil and the change of suction, obtained the LC yield surface function and hardening law of unsaturated soil, and extended the average soil skeleton stress to obtain the elliptic yield function of triaxial stress state. The quantitative parameters of soil structure proposed by Xie Jiedu et al. (2000) [74] introduced the deformation constitutive relation and strength constitutive relation of soil, and obtained the constitutive relation based on structural parameters to describe the basic law of soil deformation strength for the first time, which brought new characteristics to the study of soil mechanical properties. Zhang Ying and Deng Anfu (1997) [75] took Chongqing red clay as the research object. By analyzing the stress–strain relationship of Chongqing red clay under different stress path loading conditions, four kinds of constitutive models of Chongqing red clay were verified by using the "constitutive relationship image program of soil". It was considered that the nonlinear elastic (Chengdu University of science and Technology) K-G model and Yin Zongze double yield model. The surface elastic–plastic model can reflect the stress–strain constitutive model of Chongqing red clay and provide the corresponding model parameters.

1.3.2 Numerical Method

1. **Damage theoretical model**

In the past, many theories and models of soil mechanics were established on the basis of laboratory tests on remolded soil, while the existing in nature and engineering practice are basically undisturbed soil. The influence of soil structure on soil mechanical properties is of great significance, and the damage theoretical model of soil is developed on this basis.

Yang Minghui et al. (2015) [76] introduced the statistical damage theory, assuming that the unsaturated soil is a combination of many micro elements whose strength is subject to Weibull random distribution, used the extended Mohr Coulomb criterion to simulate the bearing capacity of micro elements, established the statistical damage evolution equation of the unsaturated soil, and then derived the statistical damage constitutive model of the unsaturated soil. Sun Long (2014) [77] applied the statistical damage theory to the unsaturated soil based on the study of the mechanical and deformation characteristics of the unsaturated soil triaxial system, constructed the micro element strength expression by using the extended Mohr Coulomb strength criterion of the unsaturated soil, and proposed the constitutive model of the unsaturated soil based on the statistical damage theory. Jiang Mingjing et al. (2013) [78] based on the mechanical theory of rock and soil damage, by considering the micro mechanism of soil structural damage, defined the structural yield surface which is similar to the geometry of the remolded soil yield surface, introduced the damage parameters to characterize the structural damage, determined the hardening law during the loading process of the structural soil, and established the constitutive model of the structural soil. Zhang Lezhong and He Qingfeng (2012) [79] based on the routine triaxial test of Yan'an Q2 loess under different confining pressure and water content, and considering the influence of damage threshold value on the damage variable of Yan'an Q2 loess, established the statistical damage constitutive model by using the statistical damage theory of rock. Xie Xing et al. (2008) [80] based on the uniaxial unconfined compression test of Q2 loess with different water content in Xi'an area, according to the Weibull statistical distribution theory and the hypothesis principle of equivalent effect change, and considering the damage threshold value of loess, deduced the damage constitutive equation of loess under the condition of uniaxial compression. Li Dongwei et al. (2007) [81] proposed the viscoplastic damage variable of frozen clay subject to DP yield criterion based on frozen soil test, and derived the viscoelastoplastic damage coupling constitutive equation of frozen soil under the related flow rule.

Generally speaking, using damage mechanics to analyze the deformation and failure of structural soil conforms to the process of people's understanding of things from macroscopic to microcosmic, from phenomenon to essence. The damage mechanics model and damage evolution equation have also successfully explained the process of deformation and failure of structural soil on the meso level. However, there are still some bottlenecks that restrict the development and engineering application of structural soil damage mechanics. How to apply structural soil damage theory to practical engineering is still a research topic of many scholars.

2. **Neural network model**

In recent years, with the development of nonlinear science, a large number of nonlinear theories have been introduced into the study of soil constitutive model. Neural network has the ability of highly nonlinear mapping, large-scale information processing and self-learning. Some scholars use neural network to carry out the numerical modeling of soil constitutive relationship.

Wang Chenghua et al. (2013) [82] established a BP neural network constitutive model of coarse-grained soil with average principal stress P and generalized shear stress Q as network input vector and body strain and shear strain as network output vector by optimizing network structure and large-scale triaxial consolidation drainage shear test of coarse-grained soil using improved BP neural network algorithm. Chen Changfu et al. (2008) [83] obtained the stress–strain relationship of grass-roots reinforced soil by the indoor triaxial test method, and established the BP neural network constitutive model of plain soil and mixed grass-roots reinforced soil based on the test results. The model has good generalization ability and can fully reflect the nonlinear relationship of geotechnical materials. Lin Fusheng (2007) [84] studied the specific implementation process of the neural network method to simulate the constitutive relationship of coarse-grained soil, and discussed the convergence of the network model curve and the extension of the network model.

Sadoun (2001) [85] uses radial basis function to estimate solar radiation. Szecsi (1999) [86] uses neural network to simulate ultrasonic sensor. Lau et al. (2000) [87] established the neural network model of the amplifier. Kong et al. (2000) [88] improved die parameters by using neural network. Qian et al. (2002), Li et al. (2002) [89, 90] introduced the application of neural network in materials science. Kim et al. (2001) [91] combined genetic algorithm and neural network method to simulate the nonlinear grinding process of cement. Chassiakost, Masri (1996) and Alehossein (2001) [92, 93] introduced the application of neural network in structural engineering. Shahin et al. (2001), Sidarta and Ghaboussi (1998) [94, 95] described the application of neural network in geotechnical and mining engineering. Ellis et al. (1995) [96] used the nested adaptive neural network to train the test data of non-uniform materials, and applied the trained model to the finite element analysis to predict the stress–strain relationship of sand. Shin and Pande (2000) [97] carried out undrained triaxial compression test of sand, and trained the stress–strain relationship of sand with different grain size grading and stress history by using neural network. Wang Jingtao (2002) [98] introduced the application of neural network in concrete.

Ren Qingyang and Wang Jingtao (2005b, 2006) [99, 100] used neural network method to establish the numerical constitutive models of expansive soil and sandy soil respectively. Zhou Baochun (2008) [101] established a numerical model of constitutive relationship of clay under different stress paths. Cheng Tao et al. (2009) [102] studied the numerical modeling of clay under different consolidation conditions.

Li Qinglai (2001) [103] used neural network method to establish a nonlinear constitutive model of clay. Luo Yidao et al. (2001) [104] applied the theory of artificial neural network and probability method to the construction of deep foundation excavation information, and compiled the deifc system of deep foundation excavation information construction software with Borland C++ Builder 5.0.

Wang Jingtao (2002) introduced the artificial neural network into the actual geotechnical engineering, established the appropriate artificial neural network model, analyzed the measured data in the dynamic compaction high fill engineering of the 11th bid section of Yuncheng Sanmenxia expressway, and predicted the shear strength of unsaturated soil with the simple and easy obtained real data such as water content and density.

She Yuexin et al. (2003) [105] proposed for the first time an elastic–plastic model for establishing the constitutive relation of rock and soil by using numerical method. He believed that the establishment of the constitutive relation of rock and soil is essentially an inverse problem, which can be directly determined by the triaxial test data, rather than by looking for the solution expression of the plastic potential.

Sun Jun et al. (2003) [106] analyzed the generalized regression neural network (GRNN), derived the relationship between the parameters of the network and the parameters of the square index correlation function, and proposed the method of calculating the correlation distance and fitting the correlation curve based on GRNN.

He Xiang et al. (2003) [107] based on the application of soft science theory to the deformation prediction and analysis of geotechnical engineering problems, combined with the construction deformation monitoring and prediction practice of the North Anchorage Foundation of Runyang Yangtze River Highway suspension bridge, carried out the multi-step rolling prediction research of the artificial intelligence neural network for the deformation of the deep and large anchorage foundation pit in the site system, which ensured the construction of the project Safety and environmental maintenance of.

Zhang Xueyan (1993) [108] elaborated the limitations of the classical slope stability analysis method, comprehensively considered the factors affecting the slope stability, and established the slope stability prediction method based on the artificial neural network. Genetic algorithm is used to optimize the structure of neural network to improve its nonlinear mapping ability and generalization ability, so as to improve the prediction accuracy. Based on the neural network trained by the existing engineering examples, the new slope stability problem is predicted.

Neural network model is built on the basis of current test data, which makes it have strong pertinence, but it is a kind of fuzzy analysis, without introducing basic mechanical equations and axioms, and may bring about calculation instability. In addition, in order to apply the neural network model to the analysis of geotechnical engineering, it is necessary to collect a large number of data to establish a sample database, that is, to do a large number of stress path tests, because the integrity and rationality of the sample database is the key to the success of model building, which limits the application of the neural network model in practical engineering.

3. **Constitutive model established by other numerical methods**

Wu Mengxi et al. (2017) [109] conducted triaxial and lateral compression tests based on Duncan E-B model, studied the influence of particle loss on the stress–strain relationship, established the quantitative relationship expression between model parameters and particle loss, and thus realized the quantitative description method of the impact of particle loss on the stress–strain relationship. Dong Qipeng et al. (2014) [110] used mathematical statistical method to study the multi-scale relationship between coarse-grained soil fabric, contact force, branch vector and coarse-grained soil fabric with direction and size, and constructed the expression of coarse-grained soil micro parameter fabric, contact force, branch vector and coarse-grained soil stress through probability statistics, least square method, orthogonal decomposition and other methods.

1.3.3 Research on the Constitutive Relationship of Expansive Soil

1. **Influence of water content on strength of expansive soil**

The expansive soil in nature is often in unsaturated state, and it has the properties of water absorption expansion, water loss shrinkage and repeated deformation. The strength of expansive soil has a direct impact on the safety of the project, which is the characterization of the ability of expansive soil to resist shear failure. Therefore, the strength index is an important parameter in the engineering design of expansive soil area. Many scholars at home and abroad have done a lot of work on the strength of expansive soil, and made a lot of valuable research results. Most of the researches on the strength of unsaturated expansive soil take Mohr Coulomb criterion as the failure criterion. Based on Bishop formula and Fredlund formula, the test data are deduced.

Chaney et al. (1996) [111] considered that the unsaturated shear strength is the result of the interaction of the strength parameters of saturated soil and the matric suction, and expressed the unsaturated shear strength by Fredlund formula. Bruyn and Thimus (1996) [112] conducted triaxial shear tests on clay under different confining pressures and temperatures, and obtained the change rule of shear strength with temperature. Ajdari et al. (2010) [113] studied the shear strength characteristics of expansive soil through the soil water characteristic curve. Miao Linchang et al. (1999) [114] conducted an experimental study on the remolded samples of expansive soil, and considered that the water content is the key factor affecting the strength, and the larger the water content is, the lower the strength is. Zhan Liangtong and Wu Hongwei (2007) [115] studied the influence of suction on the shear strength of expansive soil by controlling suction direct shear test; Kong Lingwei et al. (2010) [116] tested the deformation and strength characteristics of the remolded expansive soil in Jingmen with unsaturated soil triaxial apparatus, and found that the strength is ten points sensitive to the wetting effect. Xu Bin et al. (2011) [117] through direct shear test and triaxial test on remolded expansive soil, thought that for the same expansive soil, water content, density and fracture are three factors that affect the strength of expansive soil, among which water content and fracture have a greater impact on the strength, while density has a smaller impact on the strength. Zou Weili et al. (2012), Xiao Jie et al. (2013) [118, 119] through the shear strength test of remolded expansive soil, found that the strength parameters measured under low stress conditions are reasonable for the stability analysis of expansive soil slope. Zhang Lianjie et al. (2015) [120] analyzed the influence of water content and overburden pressure on the shear strength of Yanji remolded expansive soil.

Aitchison (1965) [121] proposed a moisture diffusion equation (shch) to describe the volume expansion and contraction deformation of expansive soil, which can predict the volume strain caused by moisture change caused by water absorption. The successful application of this model in the United States, Saudi Arabia and Australia shows that it is effective.

Bratton et al. (1990, 1991) [122, 123] discussed the influence of the humidity change of deep expansive soil on the actual project and the influence factors of the expansion and contraction deformation prediction of expansive soil subgrade under the concrete exposed surface. Dhowian et al. (1985) [124] monitored the change of moisture content of expansive soil in the field and put forward the change trend of moisture content. El garhy (1999) [125] described the influence of water change on the stability of expansive soil foundation through the settlement observation. Fredlund et al. (1993, 1997, 1997) [126–128] analyzed the influence of suction on the deformation of expansive soil raft foundation, and put forward different constitutive equations of expansive soil considering the influence of humidity. Lee and Wray (1995) [129] introduced the mechanism of soil and water action in expansive soil and the treatment of water. Lytton (1995), McKeen et al. (1976, 1980, 1990), Mitchell (1979), Pufahl and Lytton (1992) [130–135] proposed several models to predict the change of moisture content of expansive soil with seasons and other factors. Richards (1992, 1997), Ridley and Wray (1995), Russam and Coleman (1991), Wray (1989, 1995, 1998) and Mohandes (2000) [136–143] studied the suction in expansive soil respectively, and proposed several constitutive models of suction control.

2. **Study on the expansion characteristics of expansive soil**

The research on the expansion characteristics of expansive soil is mainly through the expansion force test and expansion rate test, and the expansion rate test is divided into the loaded expansion rate test and the unloaded expansion rate test.

Puffala et al. (2006) [144] studied the expansion and contraction properties of expansive soil according to the soil water characteristic curve. Erzin and Erol (2007) [145] obtained the relationship curve between matrix suction and expansion force by means of experiment and regression analysis. Xu Yongfu and Shi Chunle (1997) [146] took Ningxia expansive soil as test soil samples and carried out a large number of expansion deformation tests, and obtained the relationship between expansion rate and water content, vertical load and dry density. Yao Huayan et al. (2010) [147] selected the expansive soil of a certain engineering section to do the indoor no-load expansion rate test, studied the expansion time history characteristics of the expansive soil under the side limit condition, and obtained the empirical relationship between the no-load expansion rate of the expansive soil and the initial moisture content and compactness. Hu Jin et al. (2011) [148] studied the expansion rate of undisturbed expansive soil in the Daoxu reach of Wushen navigation through the indoor no-load expansion rate test, and divided the no-load expansion rate and time change curve into three stages: accelerated expansion stage, constant expansion stage and stable stage. Xin Zhiyu et al. (2014) [149] carried out the no-load expansion rate test of expansive soil in a construction site of Hefei under different initial moisture content conditions. The relationship between the expansion rate and the initial moisture content was fitted by using the 4-parameter Weibull model, and the relationship between the expansion index of expansive soil and the initial moisture content of soil sample was analyzed. Zhou Baochun et al. (2014) [150] took Jingmen weak expansion rate under different compactness as the sample, studied the characteristics of expansion time history curve under no-load condition, and fitted the curve with 3-parameter logistic function.

3. **Study on seepage theory of expansive soil**

The study of seepage in expansive soil is a very important aspect, but it has been ignored for a long time in the development of traditional soil mechanics. With more and more geotechnical engineering encountering the seepage problem of unsaturated soil, such as the stability analysis of embankment under the condition of water level rising and falling, the sliding failure of soil slope after rainfall, the research of seepage control measures of foundation pit and the transport of pollutants in the soil, the research on this aspect is gradually deepening.

Li Jiwei et al. (2014) [151] proposed that the study of unsaturated seepage mainly includes the occurrence and migration of water in unsaturated state and the hydraulic properties of rock and soil. In practical research, the unsaturated seepage area is constantly changing under the influence of external conditions, and the location of the seepage surface, the strength index, saturation and permeability coefficient of the soil in the seepage field will also change, which brings some difficulties to the analysis of the unsaturated seepage problem.

Lam et al. (1987) [152] put forward the saturated unsaturated seepage equation, and used the finite element method to simulate the complex transient seepage problem. Peters and Ratcliffe (1998) and Montrasio (2009) [153, 154] combined the finite element analysis of saturated unsaturated seepage with the limit equilibrium analysis method to comprehensively consider the change of air pressure, and analyzed the stability of the slope.

Zhang Peiwen et al. (2003) [155] systematically introduced the numerical simulation of saturated unsaturated unsteady flow, simulated the change of matrix suction in the slope during the rainfall process, effectively dealt with the rainfall boundary, and provided a quantitative analysis method for the stability study of rainfall type landslide. Zhang Guangcheng et al. (2007) [156] developed saturated unsaturated seepage program with APDL language in ANSYS software, and analyzed the influence of reservoir water level drop on landslide stability. Niu Wenjie et al. (2009) [157] calculated the safety factor of dam slope by considering unsaturated seepage, which is larger than that calculated only considering saturated seepage. Li Yi et al. (2012) [158] calculated the saturated unsaturated flow field through the secondary development of FLAC3D, and demonstrated the effectiveness and feasibility of this method.

1.4 Question Raising

From the previous summary, it is not difficult to see that the traditional soil constitutive models are mostly based on the test data according to different theoretical assumptions. There are two main problems in this traditional calculation mode: first, if it is easy to model and solve the parameters, it is necessary to make the model simple, but it can not meet the needs of reflecting complex engineering problems. For example, Duncan Chang model only uses the information of the initial elastic modulus and the failure stress, and can not reflect the influence of clay dilatancy, compression and stress path. Second, in order to accurately reflect the constitutive relationship of soil, the model must contain a large number of parameters. Relying on the traditional calculation mode, only a large number of assumptions can be introduced to solve these parameters. This brings many limitations to the application of the model. In the final analysis, it is caused by the limited ability of traditional computing mode to extract information.

To sum up, numerical methods such as artificial neural network have opened up a new way for the study of related problems. Using the numerical method to establish the constitutive model successfully avoids the process of artificial looking for the plastic potential function for mathematical modeling, and eliminates the use of various empirical assumptions for the convenience of determining parameters, thus greatly improving the simplicity and applicability of the model. Moreover, because of using most of the experimental data to train the network, rather than only using a few data points to determine the parameters of the model, it is obvious that the numerical modeling method has greater fault tolerance than the traditional modeling method.

In the traditional constitutive model, the relationship between constitutive relation, initial condition and stress path is not considered. However, the test results show that the influence of initial condition and stress path on the deformation of foundation can not be ignored. It has been shown that initial conditions such as water content have an important influence on the deformation of expansive soil. At the same time, Lambe (1967) [159] analyzed the relationship between the settlement and the stress path of the tank foundation, and pointed out that the strain in the vertical direction is not only related to the load size, but also to the load path. However, the influence of initial conditions and stress paths on the constitutive relationship can not be realized by the traditional calculation mode. The method of extracting information by neural network is qualitatively different from the traditional theory, and it can introduce the factor of stress path into the calculation model, and express the influence energetically. It can be said that this constitutive model, which can reflect the influence of initial conditions and stress paths, has more practical engineering significance.

1.5 Main Research Work

The first part: Taking expansive soil as the test object, triaxial compression test is carried out under different water content and drainage conditions, and network training samples are obtained. The dual yield surface model is selected as the basic framework of the constitutive model, and the elastoplastic constitutive equation of expansive soil with different water content is established. The trained model is embedded in the finite element program, and the strain response of the test expansive soil sample is calculated to verify the correctness of the model. The stress–strain curves of two kinds of expansive soil with different water content in the whole stress field (P, q) are drawn by visualization. Through comparison, it is found that there are obvious differences between them, which shows that the influence of water content on the constitutive relationship of expansive soil is quite obvious.

The second part: take sand as the test object, design three kinds of common stress paths in the project, and obtain the network training samples through the triaxial test of sand. As the same as the expansive soil model, the double yield surface model is selected as the basic framework of the constitutive model, and the elastic–plastic constitutive equation of sand under three stress paths is established. The trained equation is embedded in the finite element program, and the strain response of the test sand sample is calculated to verify the correctness of the model. At the same time, through visualization, the stress–strain relation surface and yield locus in the whole stress field (P, q) are drawn. Through the comparison of stress–strain relation surface and yield locus under the three paths, it is found that there are significant differences among them, which proves that the influence of stress path on the constitutive relation cannot be ignored.

In the third part, the elastic–plastic constitutive model of clay under the four stress paths is established, and the stress–strain relationship in the whole stress field is given, which is visualized as the spatial strain surface in the stress field. Especially, the elastic–plastic constitutive model of clay under the condition of p-reduction path and undrained condition is established, which provides a practical constitutive model for soil excavation engineering and corresponding soil engineering under undrained condition. Through the visualization of stress–strain relationship and numerical simulation, it shows that the numerical modeling method can describe the deformation characteristics of soil under different stress paths comprehensively. In addition, from the theoretical basis, basic framework and prediction results of the model, the numerical modeling method is compared with the traditional method, which shows that the numerical modeling method is superior to the traditional modeling method.

The book is divided into eight chapters. Current chapter is the introduction, which mainly describes the application of geotechnical constitutive relation and numerical modeling theory in geotechnical engineering in recent years at home and abroad, the research background, reason and main research work of this book. In Chap. 2, the stress–strain theory of soil is introduced, and the constitutive models of general soil and expansive soil are briefly introduced. Chapter 3 focuses on the numerical modeling method and its basic inverse problem theory, briefly introduces the concept, characteristics and algorithm of neural network, and compares the advantages and disadvantages of radial basis function and BP neural network. In Chap. 4, the triaxial tests of expansive soil with different water content and drainage conditions are described, as well as the establishment and verification of numerical model. In Chapter 5, the triaxial test process of sand under three stress paths is described, and the establishment and verification of numerical model are presented. In Chap. 6, the triaxial tests and numerical modeling of clay under four stress paths are described. Chapter 7 summarizes the influence of stress path and stress history on soil constitutive relation, and the influence of stress path on effective shear parameters of clay. Chapter 8 summarizes the research work of this book and looks forward to the future research.

References

1. Roscoe KH (1968) On the generalised stress-strain behaviour of wet clay. Eng Plast: 535–609
2. Mita KA, Dasari GR, Lo KW (2004) Performance of a three-dimensional hvorslev-modified cam clay model for overconsolidated clay. Int J Geomech 4(4):296–309
3. Woodward PK, Berenji AP (2001) Advanced numerical investigation of Terzaghi's super position theory. Adv Eng Softw 32(10):797–804
4. Oettl G, Stark RF, Hofstetter G (1998) A comparison of elastic-plastic soil models for 2D FE analyses of tunnelling. Comput Geotech 23(1/2):19–38
5. Lade PV, Inel S (1997) Rotaton kinematic hardening model for sand. Part 1 concept of rotating yield and plastic potential surfaces. Comput Geotech 21(3):183–216
6. Yao Y, Hou W, Luo T (2009) Unified hardening model of soil. J Rock Mech Eng 28(10):2135–2151
7. Yao Y, Niu L, Cui W et al (2011) Constitutive relation of overconsolidated unsaturated soil. Chin J Geotech Eng 33(6):833–839
8. Yao Y (2015) Research on UH model series. Chin J Geotech Eng 37(12):193–217
9. Chen B, Sun D, Jin P (2015) Study on elastoplastic constitutive model of marine sedimentary soft clay. Rock Soil Mech 36(3):730–738
10. Yao Y, Yu Y (2011) The constitutive model of sand critical state based on unified hardening parameters. Chin J Geotech Eng 33(12):1827–1832
11. Luo T, Hou W, Yao Y (2010) Constitutive model of overconsolidated soil considering asymptotic behavior. Rock Soil Mech 31(3):683–694
12. Hou W, Yao Y, Niulei (2008) Unified hardening model of structural overconsolidated soil. Indust Constr 38(8):14–16
13. Yang G, Jie Y (1999) Multiple potential surface model of soil and its verification. Chin J Geotech Eng 21(5):578–582
14. Wen Y (2014) Constitutive model of soil based on generalized potential theory and its preliminary application. Wuhan University, Wuhan

15. Tang S (2013) Study on constitutive model of peat soil in Jiangyuan town of Dunhua based on generalized potential theory. Jilin University, Changchun
16. Yao J (2010) Study on constitutive model of soil based on generalized potential theory. Wuhan University, Wuhan
17. Zhou A, Lu T (2008) The elastoplastic constitutive model of contact surface based on generalized potential theory. Chin J Geotech Eng 30(10):1532–1536
18. Yang C, Sheng D, Carter JP et al (2015) Modelling the plastic anisotropy of lower Cromer Till. Comput Geotech 69:22–37
19. Sang IW, Salgado R (2015) Bounding surface modeling of sand with consideration of fabric and its evolution during monotonic shearing. Int J Solids Struct 63:277–288
20. Dafalias Y, Taiebat M (2013) Anatomy of rotational hardening in clay plasticity. Geotechnique 63(16):1406–1418
21. Sun DA, Matsuoka H, YaoY P et al (2004) An anisotropic hardening elastoplastic model for clays and sands and its application to FE analysis. Comput Geotech 31(1):37–46
22. Manzari MT, Dafalias YF (1997) A critical state two-surface plasticity model for sands. Geomech Abstr 276(5):294–305
23. Chen Y (2016) Constitutive simulation of over consolidated soil based on thermodynamics theory. Zhejiang University, Hangzhou
24. Wang L, Shen K (2008) Study on rotational hardening law of K0 consolidated structural soft clay. Chin J Geotech Eng 30(6):863–872
25. Kong L, Hui ZX, Wang YC (2008) Application of rotational hardening and unified hardening parameters in modified Cambridge model. J Ningxia Univ 29(2):139–143
26. Wei X, Huang M (2007) Anisotropic critical state model of soft clay. Rock Soil Mech 28(9):1811–1816
27. Jie Y, Hu T, Li G et al (2004) Comparison of the multiple potential surface model and Duncan Zhang model under plane strain. Eng Mech 21(1):148–152
28. Wang J, Yao M (1996) Elastoplastic simulation of undrained cyclic behavior of soft clay. Chin J Geotech Eng 18(3):11–18
29. Huang W, Pu J, Chen Y et al (1981) Hardening law and yield function of soil. Chin J Geotech Eng 3(3):19–25
30. Xu R, Yang L, Gong X (1996) The boundary stress-strain constitutive relation of soil. J Tongji Univ 24(6):672–676
31. Oka F, Higo Y, Kimoto S (2002) Effect of dilatancy on the strain localization of water-saturated elasto-viscoplastics oil. Int Jo Solids Struct 39(13/14):3625–3647
32. Thomas HR, Cleall PJ (1999) Inclusion of expansive clay behaviour in coupled thermo hydraulic mechanical models. Eng Geol 54(1/2):93–108
33. Rosa UA, Wulfsohn D (1999) Constitutive model for high speed tillage using narrow tools. J Terra Mech 36(4):221–234
34. Stallebrass SE, Taylor RN (1997) The development and evaluation of a constitutive model for the prediction of ground movement in overconsolidated clay. Geotechnique 47(2):235–253
35. Katti DR, Desai CS (1996) Modeling and testing of cohesive soil using disturbed-state concept. Int J Rock Mech Mining Sci Geomech Abstr 33(3):114–130
36. Kong L, Yao Y (2015) Thermoviscoelastoplastic constitutive relation of overconsolidated soil. Rock Soil Mech (s1):1–8
37. Ma T, Wei C, Chen P et al (2014) Hydraulic coupling constitutive model of unsaturated soil considering residual air content. Rock Soil Mech 35(12):3415–3420
38. Liu Y, Wei C, Fang Q et al (2014) Water mechanical constitutive model of unsaturated soil and its implicit integration algorithm. Rock Soil Mech 2:365–370
39. Jin X, Zhao C (2011) Study on constitutive model of unsaturated undisturbed soil. J Beijing Univ Technol 37(1):85–91
40. Wang N (2010) The study of unsaturated soil constitutive model based on generalized effective stress. Beijing Jiaotong University, Beijing
41. Yao Y, Zhu E (2010) The method of establishing soil constitutive model based on coupling stress. Chin J Geotech Eng 32(12):1922–1929

42. Shively HL (2015) A state dependent constitutive model for rockfill materials. Int J Geomech 15(5):969–970
43. Sheng DS, Fredlund DG, Gens AG (2008) A new modelling approach for unsaturated soils using independent stress. Can Geotech J 45(4):511–534
44. Dafalias YF, Manzari MT (2004) Simple plasticity sand model accounting for fabric change effects. J Eng Mech 130(6):622–634
45. Calvello M, Finno RJ (2004) Selecting parameters to optimize in model calibration by inverse analysis. Comput Geotech 31(5):410–424
46. Ensan MN, Shahrour I (2003) A macroscopic constitutive law for elasto-plastic multilayered materials with imperfect interfaces: application to reinforced soils. Comput Geotech 30(4):339–345
47. Li J, Ding DW (2002) Nonlinear elastic behavior of fiber-reinforced soil under cyclic loading. Soil Dyn Earthq Eng 22(4):977–983
48. Dafalias YF, Manzari MT, Papadimitriou AG (2002) A simple anisotropic clay plasticity model. Mech Res Commun 29(4):241–245
49. Zhang HW, Sanavia L, Schrefler BA (2001) Numericla analysis of dynamic strain localisation in initially water saturated dense sand with a modified generalised plasticity model. Comput Struct 79:441–459
50. Wan RG, Guo PJ (1998) A simple constitutive model for granular soils: modified stress-dilatancy approach. Comput Geotech 22(2):109–133
51. Boulon M, Alachaher A (1995) A new incrementally nonlinear constitutive law for finite element application in geotechnnics. Comput Geothnics 17:177–201
52. Chowdhury EQ, Nakai T, Tawada M et al (1999) A model for clay using modified stress under various loading conditions with the application on subloading concept. Soils Foud Jpn Geotech Soc 39(6):103–116
53. Wheeler SJ, Sivakumar V (1995) An elasto-plastic critical state framework for unsaturated soil. Geotechnique 45(1):35–53
54. Goto S, Burland JB, Tatsuoka F (1999) Nonlinear soil model with various strain levels and its application to axisymmetric excavation problem. Soil Found Jpn Geotech Soc 39(4):111–119
55. Ortiz M, Pandolfi A (1999) A variational cam-clay theory of plasticity. Comput Methods Appl Mech Eng 193(27):645–662
56. Li X, Thomas HR, Fan Y (1999) Finite element method and constitutive modelling and computation for unsaturated soils. Comput Method Appl Mech Eng 169(1/2):356–367
57. Adachi T, Oka F, Zhang F (1998) Anelasto-viscoplasticconstitutive model with strain softening. Soil Fund Jpn Geotech Soc 2:27–35
58. Krogsboll A (1998) Constitutive model with time-dependent deformations. Eng Geol 49(3):285–292
59. Fox PJ, Berles JD (1997) A piecewise-linear model for large strain consolidation. Geomech Abstr 276(5):306–320
60. Niemunis A, Krieg S (1996) Viscous behaviour of soil under oedometric conditions. Int J Rock Mech Mining Sci Geomech Abstr 33(8):356–370
61. Bardet JP (1996) Scaled memory model for cyclic behavior of soils. Int J Rock Mech Mining Sci Geomech Abstr 33(5):203–223
62. Leidwanger-Rabis CA, Chambon RA, Catel PB et al (1995) A parametric analysis for the Dafalias-Kaliakin bounding surface viscoplastic model based on the Diflupress L D. test. Comput Geotech 17(4):473–505
63. Duncan JM, Chang CY (1970) Nonlinear analysis of stress and strain in soils. J Soil Mech Found Div 96(SM5):1629–1653
64. Wei G, Chen S, Li G et al (2016) Constitutive model of coarse grained soil for dam construction based on state parameters. Chin J Geotech Eng 38(4):654–661
65. Zhang B, Zhu S, Li Y (2015) Generalized plastic constitutive model of overconsolidated soil. Water Resour Power 8:112–115
66. Li W, Yang Q, Liu W et al (2015) Study on constitutive model of unsaturated expansive soil based on SFG model. Chin J Geotech Eng 37(8):1449–1453

67. Liu Y, Wei C, Zhao C et al (2013) A constitutive model of unsaturated soil with high saturation. Rock Soil Mech 8:2189–2194
68. Liu M (2011) The modification and research of SFG model. Beijing Jiaotong University, Beijing
69. Jia Y (2010) Constitutive relation of coarse grained soil considering particle breakage effect. Rock Soil Mech 31(7):2111–2115
70. Jia Y, Chi S, Lin G (2010) Unified dilatancy constitutive model of coarse grained soil considering particle breakage. Rock Soil Mech 31(5):1381–1388
71. Sun H, Huang M (2009) Constitutive model considering strain softening and dilatancy of coarse grained soil. J Tongji Univ 37(6):727–732
72. Li G, Si W, Zhang Q (2008) Tsinghua elastoplastic model of unsaturated soil. Rock Soil Mech 29(8):2033–2036
73. Miao L (2007) Study on constitutive model of unsaturated soil. Rock Soil Mech 28(5):855–860
74. Xie J, Qi J, Zhang Z (2000) Considering the constitutive relationship of soil structure. Chin Civil Eng J 33(40):35–41
75. Zhang Y, Deng A (1997) Verification of constitutive model of Chongqing red clay. J Chongqing Jianzhu Univ (Social Science Edition) 19(2):48–53
76. Yang M, Sun L, Zhao M et al (2015) Simple constitutive model of unsaturated soil based on statistical damage theory. Hydrogeol Eng Geol 42(3):43–48
77. Sun L (2014) Experimental study and engineering application of statistical damage constitutive model of unsaturated soil. Hunan University, Changsha
78. Jiang M, Sun Y, Zhang F (2013) Discrete element simulation of damage law of cemented geotechnical materials based on micromechanics. Rock Soil Mech 7:2043–2050
79. Zhang L, He Q (2012) Study on statistical damage constitutive model of Yan'an Q2 loess. J Eng Geol 20(6):1064–1070
80. Xie X, Wang D, Zhao F (2008) Research on damage constitutive model of structural Q2 loess under uniaxial compression. Hydrogeol Eng Geol 35(3):47–50
81. Li D, Wang R, Hu P (2007) Study on the constitutive relationship of frozen clay coupled with creep damage. J Glaciol Geocryol 29(3):446–449
82. Wang C, Sujuan LF (2013) Study on BP neural network constitutive model of coarse grained soil. Low Temp Archit Technol 35(4):128–130
83. Chen C, Pengzhao, Liu H (2008) The constitutive model of grass root reinforced soil based on BP neural network. Bull Soil Water Conserv 28(3):93–96
84. Lin F (2007) Study on neural network model of constitutive relation of coarse grained soil. Tianjin University, Tianjin
85. Sadoun B (2001) An efficient simulation scheme for testing materials in a nondestructive manner. Inf Sci 137:43–51
86. Szecsi T (1999) Cutting force modelling using artificial neural networks. J Meter Process Technol 92:344–349
87. Lau HCW, Ning A, Pun KF et al (2000) Neural network for the dimensional control of molded parts based on a reverse process model. J Meter Process Technol 117:89–96
88. Kong LX, Hodgson PD, Collinson DC (2000) Extrapolative prediction of the hot strength of austenitic steels with a combined constitutive and ANN model. J Mater Process Technol 102:84–89
89. Qian HC, Xia BC, Li SZ et al (2002) Fuzzy neural network modeling of material properties. J Mater Prosess Technol 122:196–200
90. Li SP, Bin HZ, Huang ZC et al (2002) Nonlinear comminution process modeling based on GA-FNN in the computational comminution system. J Mater Process Technol 120:84–89
91. Kim JH, Kim YH (2001) A predictor-corrector method for structural nonlinear analysis. Comput Method Appl Mech Eng 191:959–974
92. Chassiakos AG, Masri SF (1996) Identification of structure systems by neural networks. Math Comput Simul 40:637–656
93. Alehossein H (2001) Data mining for geotechnical and mining engineering. Aust Geomech: 49–54

94. Shahin MA, Jaksa MB, Maier HR (2001) Artificial neural network applications in geotechnical engineering. Austr Geotech: 49–62
95. Sidarta DE, Ghaboussi J (1998) Constitutive modeling of geomaterials from nonuniform material tests. Comput Geotech 22(1):53–71
96. Ellis GW, Yao C, Zhao R et al (1995) Stress-strain modeling of sands using artificial neural networks. J Geotech Eng 121(5):429–435
97. Shin HS, Pande GN (2000) On self-learning finite element codes based on monitored response of structure. Comput Geotech 27:161–178
98. Wang J (2002) The numerical method of establishing the constitutive model of rock and soil. J Huazhong Univ Sci Technol (Natural Science Edition) 19(1):44–47
99. Ren Q, Wang J (2005) Numerical modeling of sand elastoplastic constitutive relation under equal principal stress ratio path. J Huazhong Univ Sci Technol (Natural Science Edition) 22(3):69–71
100. Ren Q, Wang J (2006) Numerical modeling of elastoplastic constitutive relation of expansive soil. Chin J Geotech Eng 26(10):1699–1703, 1708
101. Zhou B (2008) The theory and experimental study of clay constitutive modeling. Huazhong University of science and technology, Wuhan
102. Cheng T, Yan K, Wang J (2009) Numerical modeling of clay under different consolidation conditions. Rock Soil Mech 30(11):3352–3356
103. Li Q (2001) Application of artificial neural network and probability method in information construction of deep foundation excavation. Chin Civil Eng J 34(2):96–100
104. Luo Y, Wang Z, Fan J (2001) A method for predicting shear strength of unsaturated soil. Hydropower Pumped Storage 25(6):41–44
105. She Y, Liu H, Gao Y (2003) Neural network method for calculating correlation distance. Rock Soil Mech 24(5):719–722
106. Sun J, Zhao Q, Xiong X (2003) Intelligent prediction of construction deformation of North Anchorage Foundation of Runyang Yangtze River Highway Bridge Engineering Record research. Rock Soil Mech 24:1–7
107. He X, Li S, Liu Y et al (2003) Artificial neural network method for the stability prediction of rock and soil slope. Rock Soil Mech 24:73–76
108. Zhang X (1993) Plastic mechanics of rock and soil. China Communications Press, Beijing
109. Wu M, Ye Z, Zhang Q (2017) Study on the effect of fine particle loss on the constitutive relationship of sandy gravel soil. Rock Soil Mech 38(6):1550–1556
110. Dong Q, Yao H, Lu Z et al (2014) Study on three dimensional micro fabric and constitutive relationship of coarse grained soil. Chin J Rock Mech Eng 33(2):4334–4342
111. Chaney R, Demars K, Vanapalli S et al (1996) The relationship between the soil-water characteristic curve and the unsaturated shear strength of a compacted glacial till. Geotech Test J 19(3):259–268
112. Bruyn DD, Thimus JF (1996) The influence of temperature on mechanical characteristics of boom clay: the results of an initial laboratory programme. Eng Geol 41(1/2/3/4):117–126
113. Ajdari M, Habibagahi G, Nowamooz H et al (2010) Shear strength behavior and soil water retention curve of a dual porosity silt-bentonite mixture. Scientia Iranica 17(6):430–440
114. Miao L, Zhong X, Yin Z (1999) The relationship between strength and water content of expansive soil. Rock Soil Mech 20(2):71–75
115. Zhan L, Wu H (2007) Effect of suction on shear strength and dilatancy of unsaturated expansive soil. Chin J Geotech Eng 29(1):82–87
116. Kong L, Zhou B, Bai H et al (2010) Experimental study on deformation and strength characteristics of Jingmen unsaturated expansive soil. Rock Soil Mech 31(10):3036–3042
117. Xu B, Yin Z, Liu S (2011) Experimental study on influencing factors and laws of expansive soil strength. Rock Soil Mech 32(1):44–50
118. Zou W, Chen L, Xie P et al (2012) Nonlinear strength characteristics of remolded expansive soil and one-dimensional consolidation swelling stress-strain relationship. Rock Soil Mech 33(S):59–64

119. Xiao J, Yang H, Li H et al (2013) Shear strength test of Nanning expansive soil with different density under low stress condition. China J Highway Transp 26(6):15–21
120. Zhang L, Wu X, Xie Y et al (2015) Influence of water content and overlying pressure on shear strength of remolded expansive soil. Chin J Geol Hazard Control 4:138–143
121. Aitchison GD (1965) Statement of the review panel: engineering concepts of moisture equilibria and moisture changes in soils beneath covered area. In: Proceedings moisture equilibria and moisture changes in soils beneath covered areas, Butterworths, pp 7–21
122. Bratton WL, Lee CS, Pomper PS et al (1990) Field monitoring of moisture conditions in expansive soils. In: Proceedings, symposium on engineering geology and soils engineering, pp 36/1–36/14
123. Bratton WL (1991) Parameters for predicting shrink/heave beneath slab-on-ground foundations over expansive clays. Texas Technology University, Texas
124. Dhowian AW, Erol O, Youssef AF (1985) Evaluation of expansive soils and foundation methodology in the Kingdom of Saudi Arabia. Research Report. No. AT-5-88-3, CANCST
125. El-Garhy BM (1999) Soil suction and analysis of raft foundation resting on expansive soils. Dissertation, Minufiya University, Al Minufya, Egypt
126. Fredlund DG, Rahardjo H (1993) Soil mechanics for unsaturated soils. Wiley, NewYork
127. Fredlund DG, Hung VQ (1997) Prediction of volume change in an expansive soil as a result of vegetation and environmental changes. Geotech Spec Publ 115:24–43
128. Fredlund DG (1997) An introduction to unsaturated soil mechanics. Proc Unsaturated Soil Eng Practi Geotech Spec Publ 68:1–37
129. Lee H C, Wray W K. 1995. Techniques to evaluate soil suction-A vital unsaturated soil water variable. In: 1st international conference on unsaturated soils, Paris, pp 615–622
130. Lytton RL (1995) Prediction of movement in expansive clays. Proc Vert Horiz Deform Found Embankments 2:1827–1845
131. McKeen RG (1976) Design and construction of airport pavements on expansive soils. United States Department of Transportation Report, No. FAA-RD-76-66, Washington
132. McKeen RG (1980) Field studies of airport pavements on expansive clay. In: 4th international conference on expansive soils, Denver, pp 242–261
133. McKeen RG, Johnson LD (1990) Climate-controlled soil design parameters for mat foundations. J Geotech Eng 116(7):1073–1094
134. Mitchell PW (1979) The structural analysis of footings on expansive soil. Research Report, No. 1, Adelaide
135. Pufahl DE, Lytton RL (1992) Temperature and suction profiles beneath highway pavements: Computed and measured. In: Transportation Research Record 1307, Transportation Research Board, pp 268–276
136. Richards BG (1992) An analysis of subgrade conditions at the Horsham experimental road site using two-dimensional diffusion equation on a high-speed digital computer. In: Proceedings moisture equilibria and moisture changes in soils beneath covered areas, Butterworths, pp 243–258
137. Richards BG (1997) Moisture flow and equilibria in unsaturated soils for shallow foundations. In: Permeability and capillarity of soils, ASTM No. STP417, Baltimore, pp 4–34
138. Ridley AM, Wray WK (1995) Suction measurements: a review of current theory of practices. In: 1st international conference on unsaturated soils, vol 3, pp 1293–1322
139. Russam K, Coleman JD (1991) The effect of climatic factors on subgrade moisture conditions. Geotechnique 11(1):22–28
140. Wray WK (1989) Mitigation of damage to structures supported on expansive soils. Nation Sci Found 1:1239–1250
141. Wray WK (1995) Using soil suction to estimate differential soil shrink or heave. Proc Unsaturated Soil Eng Pract 45(1):30–45
142. Wray WK (1998) Mass transfer in unsaturated soils: a review of theory and practices. In: Proceedings, 2nd international conference on unsaturated soils, vol 2, pp 99–155
143. Mohandes M, Balghonaim A, Kassas M et al (2000) Use of radial basis functions for estimating monthly mean daily solar radiation. Sol Energy 68(2):161–168

144. Puppala AJ, Punthutaecha K, Vanapalli SK (2006) Soil-water characteristic curves of stabilized expansive soils. J Geotech Geoenviron Eng 132(6):736–751
145. Erzin Y, Erol O (2007) Swell pressure prediction by suction methods. Eng Geol 92(3/4):133–145
146. Xu Y, Shi C (1997) Strength characteristics of expansive soil. J Yangtze River Sci Res Inst 14(1):38–41
147. Yao H, Wu Y, Huang B (2010) Experimental study on expansive soil expansion under confined condition. Water Resour Power 28(7):40–43
148. Hu J, Wang B, Zhang W et al (2011) Study on deformation law of expansive soil under no-load and loaded conditions. Chin J Geotech Eng 33(1):335–338
149. Xin Z, Tan X, Hu N et al (2014) Experimental study and variability analysis of swelling shrinkage index of expansive soil. J Guangxi Univ (Natural Science Edition) 39(1):124–131
150. Zhou B, Zhang Y, Tang Z et al (2014) One dimensional expansion compression characteristics of Jingmen weak expansive soil with different compactness. Rock Soil Mech 35(5):1275–1283
151. Li J, Wang H, Zhang L (2014) One dimensional analytical solution and parameter analysis of coupling stress field and seepage field of unsaturated soil. Hydrogeol Eng Geol 41(4):57–61
152. Lam L, Fredlund DG, Barbour SL (1987) Transient seepage model for saturated-unsaturated soil systems: Ageot. Can Geotech J 24(24):565–580
153. Peters NE, Ratcliffe EB (1998) Tracing hydrologic pathways using chloride at the panola mountain research watershed, Georgia, USA. Water Air Soil Pollut 105(1/2):263–275
154. Montrasio L, Valentino R, Losi GL (2009) Rainfall-induced shallow landslides: a model for the triggering mechanism of some case studies in Northern Italy. Landslides 6(3):241–251
155. Zhang P, Liu D, Huang D et al (2003) Numerical simulation of saturated-unsaturated unsteady seepage. Rock Soil Mech 24(6):927–930
156. Zhang G, Tang H, Hu B (2007) Influence of unsaturated seepage on landslide stability. Rock Soil Mech 28(5):965–970
157. Niu W, Ye W, Liu S et al (2009) Limit analysis of soil slope considering saturated unsaturated seepage. Rock Soil Mech 30(8):2477–2482
158. Li Yi Wu, Jia LK (2012) Saturated unsaturated seepage analysis based on FLAC3D. Rock Soil Mech 33(2):302–307
159. Lambe TW (1967) Stress path method. J Soil Mech Found Div 93(118):1195–1217

Chapter 2
The Constitutive Relationship of Rock and Soil and Its Influencing Factors

The central part of numerical processing of a physical problem is the description of some physical quantities, such as the relationship between stress–strain and time, which are called constitutive relations. This chapter lists some relationships which have been widely and successfully applied in geotechnical engineering. We first discuss some basic concepts that are applicable to any material.

2.1 Soil Stress–Strain

2.1.1 Stress

Equation (2.1) represents the micro-element of a larger continuum in soil. The force is transferred on six surfaces of the micro element and can be expressed as stress tensor:

$$\sigma_{\mathrm{ij}} = \begin{bmatrix} \sigma_{\mathrm{xx}} & \tau_{\mathrm{xy}} & \tau_{\mathrm{xz}} \\ \tau_{\mathrm{yx}} & \sigma_{\mathrm{yy}} & \tau_{\mathrm{yz}} \\ \tau_{\mathrm{zx}} & \tau_{zy} & \sigma_{zz} \end{bmatrix} \tag{2.1}$$

$$\tau_{\mathrm{xy}} = \tau_{\mathrm{yx}}$$
$$\tau_{\mathrm{yz}} = \tau_{zy}$$
$$\tau_{\mathrm{zx}} = \tau_{\mathrm{xz}}$$

$$p_{ij} = \begin{bmatrix} p & 0 & 0 \\ 0 & p & 0 \\ 0 & 0 & p \end{bmatrix} \tag{2.2}$$

Q. Ren and J. Zhou, *Numerical Modeling of Soil Constitutive Relationship*,
https://doi.org/10.1007/978-981-16-3231-0_2

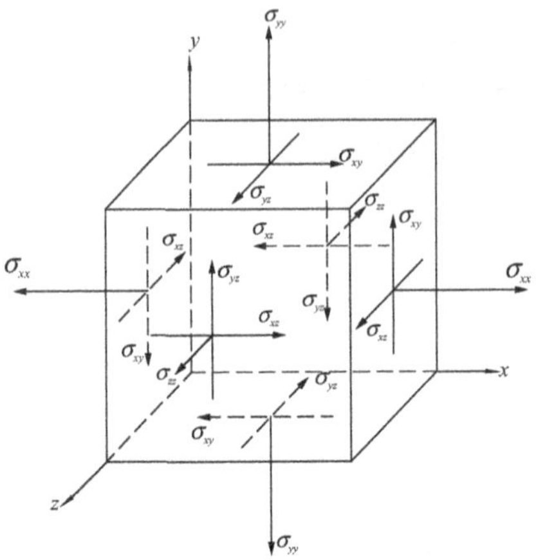

(a) The tensile stress is positive

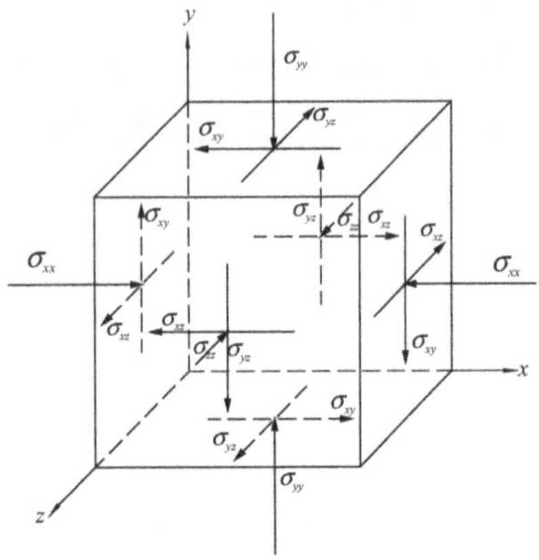

(b) The compressive stress is positive

Fig. 2.1 Soil stress

The symbol of Eq. (2.1) is commonly used in continuum mechanics. In continuum mechanics, the stress is positive when the direction of the external normal of the force and its acting surface are either positive or negative, which makes the tensile force positive, as shown in Fig. 2.1. Since pressure is more common than tension in geotechnical engineering problems, it is usually assumed that the pressure is positive. Equation (2.2) is the tensor expression of pore pressure.

If the same symbol is used for the total stress and pore pressure, the effective stress:

$$\overline{\sigma} = \sigma - p \tag{2.3}$$

or

$$\begin{bmatrix} \overline{\sigma}_{xx} & \overline{\tau}_{xy} & \overline{\tau}_{xz} \\ \overline{\tau}_{yx} & \overline{\sigma}_{yy} & \overline{\tau}_{yz} \\ \overline{\tau}_{zx} & \overline{\tau}_{zy} & \overline{\sigma}_{zz} \end{bmatrix} = \begin{bmatrix} \sigma_{xx} & \tau_{xy} & \tau_{xz} \\ \tau_{yx} & \sigma_{yy} & \tau_{yz} \\ \tau_{zx} & \tau_{zy} & \sigma_{zz} \end{bmatrix} - \begin{bmatrix} p & 0 & 0 \\ 0 & p & 0 \\ 0 & 0 & p \end{bmatrix} \tag{2.4}$$

The stresses discussed later in this book are all effective stresses.

2.1.2 Strain

Corresponding to stress, the strain of soil can be expressed as:

$$\varepsilon_{ij} = \begin{bmatrix} \varepsilon_{xx} & \gamma_{xy} & \gamma_{xz} \\ \gamma_{yx} & \varepsilon_{yy} & \gamma_{yz} \\ \gamma_{zx} & \gamma_{zy} & \varepsilon_{zz} \end{bmatrix} \tag{2.5}$$

where,

$$\gamma_{xy} = \gamma_{yx}$$
$$\gamma_{yz} = \gamma_{zy}$$
$$\gamma_{zx} = \gamma_{xz}$$

Strain is the result of stress, the change in the magnitude and direction of the stress can produce strain. Strain can be divided into elastic strain, viscous strain and plastic strain, or can be divided into volume strain and shear strain. Strain is not only affected by the properties of the material itself, but also affected by complex factors such as previous stress, current stress and stress changes.

2.2 Constitutive Relation Model of Soil

From a macro point of view, the discontinuity in the soil microstructure can be ignored. Therefore, the study of soil constitutive model belongs to the category of constitutive theory of continuum mechanics in theory. The basic equations of continuum mechanics can be used to solve the problems of soil mechanics. These equations include equilibrium equation or motion equation, compatibility condition of strain and displacement (geometric equation), constitutive equation of material. The equilibrium equation and geometric equation have nothing to do with the properties of materials, while the constitutive equations of materials are different due to the different properties of materials. Therefore, once the constitutive equation of the material is known, the response of the material medium to the external load can be determined by adding the balance equation and the coordination equation.

The models established for practical application are all formed by generalizing the real situation. These generalized models are based on a certain scope of use and basic assumptions; it is very difficult to establish a model that is completely consistent with the real situation.

In recent years, with the rapid development of soil plastic mechanics, a relatively complete system has gradually formed. Therefore, the so-called new model is only a model established on the basis of summarizing the experience of the original model and improving its shortcomings for specific problems.

The constitutive model of soil can be divided into elastic model and plastic model, and the plastic model can be divided into elastic–plastic model, viscoelastic-plastic model, internal time plastic model and so on. According to the different material parameters used in the model, the constitutive relationship of soil can be divided into *E-v* type model, *K-G* type model, λ-μ type model and so on. If the constitutive model of soil is classified from the expression mode, it can be divided into two types: full quantity type and incremental type. The full constitutive relationship is to establish the constitutive relationship between the total stress (total stress) and the total strain (total strain); the incremental constitutive relationship is to establish the constitutive relationship between stress increment ($d\sigma ij$) and strain increment ($d\varepsilon ij$).

Since the plastic constitutive relationship is related to the stress history and stress path, the plastic constitutive relationship is incremental in nature. The full quantity type is just a special case of the incremental type.

2.2.1 Elastic Model

Elastic models can be divided into linear elastic models and nonlinear elastic models.

(1) Linear elastic model. Love proposed a linear elastic model, which assumes that the stress of soil is proportional to the strain, obeys Hooke's law, and the strength is infinite. Based on this theory, the linear elastic constitutive relation can be expressed as:

$$\sigma_{\mathrm{ij}} = D^{\mathrm{e}}_{\mathrm{ijkl}}\varepsilon_{\mathrm{kl}} \tag{2.6}$$

or

$$\varepsilon_{ij} = C^{e}_{ijkl}\sigma_{kl}$$

where, D^e, C^e are stiffness matrix and flexibility matrix, respectively. For extremely anisotropic materials, they contain 81 constants; for orthotropic materials, they contain 9 constants; for transversely anisotropic materials, their constants are reduced to 5; and for isotropic materials, they contain only two constants. The linear elastic models have been widely used in geotechnical numerical calculations. But in general, this model is too simplified and has a smaller scope of application.

(2) Nonlinear elastic model. If geotechnical materials are regarded as elastic materials, the constitutive relationship is essentially closer to nonlinear elasticity, while linear elastic materials are only a simplification of stress–strain relationship of soil. Therefore, if the geotechnical materials are regarded as elastic materials, the nonlinear elastic constitutive relationship should be used to simulate it, which is more realistic. The nonlinear elastic constitutive relationship can be roughly divided into three types: Cauchy elastic model, hyper elastic model and sub elastic model. Among them, the essence of the Cauchy elastic model is that the stress (or strain) of the elastic material

depends exclusively on the current strain (or stress); The hyper elastic model is a kind of constitutive equation established by strain energy function or complementary energy function of material; The secondary elastic model is established when the stress state of elastic material is not only related to the strain state, but also to the stress path to the state. Among the above three models, the simple variable modulus sub elastic model is the most widely used.

In the study of stress–strain relationship, the nonlinear elastic analysis of soil has been studied a lot. For example, For example, Desai et al. used nonlinear analysis of ax symmetric foundation on layered clay, and Duncan et al. used the assumption that Poisson's ratio is a constant, which makes the elastic modulus change with the stress state. Clough used constant bulk deformation modulus and varying shear modulus obtained from triaxial tests to analyze the displacement of ottsebrook dam during construction. Taylor introduced the method that shear modulus is related to stress invariant. On the basis of these studies, some nonlinear models of rock and soil are gradually established, for example, the nonlinear sub elastic model proposed by Duncan and Zhang Jinrong (also known as Duncan Chang model); The sub elastic and variable modulus model (I-K model for short) that considers the shear expansion of rock and soil was proposed by Izumi and Kamemura. Among the existing nonlinear models, the Duncan Chang model is widely used and representative.

The Duncan-Chang model assumes that the stress–strain relationship is hyperbolic, so it is also called the hyperbolic model, as shown in Fig. 2.2. In the Fig. 2.2, Ei is the initial elastic modulus, $(\sigma 1-\sigma 3)_{ua}$ is the maximum principal stress difference; εa is the axial strain, and $\varepsilon a = \varepsilon 1$.

Among them, the tangent modulus Et and Poisson's ratio vt vary with the stress level. As long as Et and vt of different stress levels are reasonably determined through experiments and calculations, the stress–strain analysis and calculation can be carried out according to the incremental generalized Hooke's law. The mathematical expression of the Duncan-Chang model is very simple and the concept is clear; there are only 8 experimental constants, and the physical and geometric meanings are clear; To a certain extent, it also reflects the nonlinear elasticity of geotechnical materials and its characteristics related to stress path, so it is widely used in engineering and has

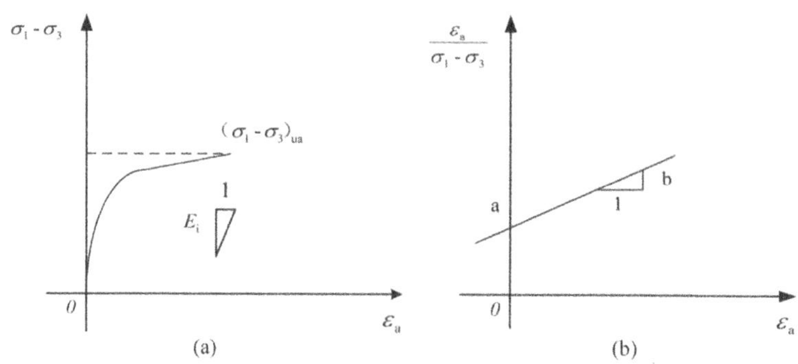

Fig. 2.2 Duncan Chang model

accumulated a lot of experience. However, its disadvantage is that it does not consider the dilatancy and compressive rigidity of geotechnical materials, and the influence of intermediate principal stress on deformation and strength. In view of the above shortcomings of Duncan Chang model, scholars at home and abroad have proposed many improvements. Skermer proposed a nonlinear model considering the effect of intermediate principal stress, in which the elastic modulus and Poisson's ratio are changed according to the variation of confining pressure and deviator stress. The confining pressure here is assumed to be the mean value of the small principal stress and the medium principal stress, and the deviator stress is the difference between the large principal stress and the confining pressure. Shen Zhujiang [1] treated the volume strain caused by shear dilatancy as initial strain, changed the $q-\varepsilon 1$ curve into strain softening curve with hump, and proposed a nonlinear model with $Mt - Et - \rho$ three parameters. Where Mt is the tangent compression modulus and ρ is the shear expansion coefficient.

In order to make the E-V model reflect the change of certain stress path, Yudhbir established a nonlinear model based on the triaxial test results and elastic theory, which can reflect the changes of E and ν under different stress paths. Schulhse and Teosen proposed a model that describes the principal stress ratio K as a constant stress path based on the E-v model. Liu Zude and others in China use power function to describe the stress–strain relationship. Using the data obtained from the undisturbed London clay test, Wroth concluded that the parameters K and G of the soil elastic model are mainly determined by P', and the overconsolidation ratio has a relatively small effect on them.

The $K - G$ model separates the spherical tensor component and the partial tensor component of stress and strain. The bulk modulus K is used to reflect the volume change of soil, and the shear modulus G is used to reflect the shear deformation of soil. In recent years, more and more researches have been done on $K - G$ model because of its clear physical meaning and easy connection with stress. Domaschuk and Valliappan suggested that bulk modulus K and shear modulus G should be used instead of E and υ commonly used in engineering when performing nonlinear elastic analysis. Assuming that the soil is isotropic and the stress–strain directions coincide, the average stress and volumetric strain are related to the volumetric deformation modulus K as follows: $\sigma m = K\varepsilon \upsilon$, while the relationship between stress Deviator and strain deviator by shear modulus is as follows: $S'ij = 2G\varepsilon' ij$. On this basis, the Domaschuk-Valliappan model was established. Naylor believes that the bulk deformation modulus increases with the increase of the average normal stress; while the shear modulus increases with the increase of the average normal stress, and decreases with the increase of the shear stress. He suggested that the expressions of K and G in nonlinear elastic analysis are:

$$K_t = K_I + \alpha_k p \tag{2.7}$$

$$G_t = G_I + \alpha_G p + B_G q \tag{2.8}$$

where, αk, GI, KI, αG and BG are all the test constants.

The above models assume that the soil is isotropic, but the real soil is mostly anisotropic. On the basis of considering both reality and simplification, the common method is to assume that the soil has the transverse isotropy or orthotropic of the symmetry axis. Graham and Houlsby proposed a matrix equation for transversely isotropic soils expressed by triaxial stress invariants:

$$\begin{bmatrix} \delta P' \\ \delta Q' \end{bmatrix} = \begin{bmatrix} K^m & J \\ J & 3G^m \end{bmatrix} \begin{bmatrix} \delta \varepsilon_v \\ \delta \varepsilon_s \end{bmatrix} \tag{2.9}$$

where, K^m, G^m and J can be obtained by triaxial test of transversely isotropic soil samples.

Shen Zhujiang [2] of the Nanjing Institute of Hydraulic Research and others expressed the stress–strain relationship of soil with the following two nonlinear functions:

$$\varepsilon_v = f_1(p, q) \tag{2.10}$$

$$\overline{\varepsilon} = f_2(p, q) \tag{2.11}$$

where, P, q, εv and $\overline{\varepsilon}$ are octahedral stresses and their corresponding strains, respectively. Rewritten as incremental as:

$$\mathrm{d}\varepsilon_v = \frac{\partial f_1}{\partial p}\mathrm{d}p + \frac{\partial f_1}{\partial q}\mathrm{d}p \tag{2.12}$$

$$\mathrm{d}\overline{\varepsilon} = \frac{\partial f_2}{\partial p}\mathrm{d}p + \frac{\partial f_2}{\partial q}\mathrm{d}p \tag{2.13}$$

Compared with the classical elastic theory, the above two formulas respectively point out: ① pure shear can produce volumetric strain; ② Changes in the sum of normal stresses can cause shear strain. This reflects the shear shrinkage and dilatancy, and negates the assumption caused by the existence of two elastic constants E and ν.

The $f1$ function can be obtained by isotropic consolidation, conventional unidirectional consolidation tests, and other consolidation tests that keep $n = q/p$ constant. The $f2$ function can be obtained by a triaxial compression drain test with p equal to a constant.

Whether linear elastic model or nonlinear elastic model, their common basic feature is that the stress and strain are reversible, or reversible in incremental sense. This model can get more accurate results when used for monotonous loading. But when it is used to solve complex loading problems, its accuracy often cannot meet the needs of engineering. The common shortcomings of this elastic model limit the application of the elastic model, so it limits the development of elastic–plastic constitutive model and its wide application in various geotechnical problems. At present, the elastoplastic constitutive model of soil based on modern plasticity theory is a more complete and widely used model.

2.2.2 Plastic Model

1. Elastoplastic model

The yield criterion is an important concept of plastic mechanics, which is the basis of judging the elastoplasticity of materials. Therefore, before discussing the elastic–plastic model, the existing yield criteria are summarized. The existing yield surfaces can be roughly divided into two types: The existing yield surfaces can be roughly divided into two types: one is a single-opening yield surface, also called a cone yield surface; the other is a closed yield surface that is currently widely used, also called a hat yield surface. The open cone yield surface mainly reflects plastic shear deformation. Most of the classical yield surfaces belong to this type, such as Tresca criterion, Mises criterion, Mohr Coulomb criterion, Drucker Prager criterion, etc. However, one of the distinguishing characteristics of geotechnical materials different from metal materials is that pure hydrostatic pressure can also produce plastic volumetric strain. The yield surface with a single opening can not reflect the plastic volumetric strain. Therefore, in recent years, whether the original yield surface is modified or the new yield surface is proposed, most of them are the hat yield surface which overcomes some shortcomings of single yield surface, and can describe the properties of soil more truly and fit the test data under various loading paths properly. It can be widely used for different purposes, and has adaptability and flexibility, so it has been widely developed and applied in recent years.

Although there are many kinds of yield surfaces, it can be found that they are convex in the two planes, and most of them take the double shear stress strength theory proposed by Yu Maohong [3] as the outer boundary, and the Mohr Coulomb single shear strength theory as the inner boundary. The shapes on the meridian surface can be roughly divided into the following types: egg shape with both ends round, water drop shape with one tip at one end and Olive shaped with two tips at both ends.

Yu Maohong [4] proposed the yield and failure criteria of double shear stress for metal materials and geotechnical materials respectively. Based on the modification of the original criterion, the yield surface of generalized double shear stress elliptical cap was proposed later. In this criterion, the effects of intermediate principal stress, Rhode angle and hydrostatic pressure on Yield and failure are considered, and different strength (S-D) effects with different tensile and compressive strengths are considered, and the edges and corners of the original model are rounded. The parameters in the criterion are easy to determine or can be converted from the material parameters of Tresca criteria and Mohr–Coulomb criteria. The criterion has clear mechanical and physical significance, and has a wide application prospect for geotechnical materials.

According to the elastoplastic theory, the total strain can be divided into two parts: elastic strain and plastic strain, and the incremental form is $d\varepsilon ij = d\varepsilon ij^{v} + d\varepsilon ij^{p}$. Among them, the elastic strain can be calculated by the generalized Hooke's law, and the plastic strain can be calculated by the plastic increment theory. It is necessary to know the yield function, flow rule and hardening law of the material to calculate the plastic strain by plastic increment theory, and the plastic potential function of the material should be known for the material subject to the unrelated flow rule. The elastoplastic constitutive equation can be expressed as:

$$\mathrm{d}\sigma_{ij} = D_{ijkl}^{ep}\mathrm{d}\varepsilon_{ijkl}$$

where, D_{ijkl}^{ep} is the elastic–plastic stiffness matrix, and its general expression is:

$$D_{ijkl}^{\mathrm{ep}} = D_{ijkl} - \frac{D_{ijpg}\frac{\partial g}{\partial \sigma_{pg}}\frac{\partial \phi}{\partial \sigma_{rs}}D_{rskl}}{A + \frac{\partial \phi}{\partial \sigma_{mn}}D_{mnuv}\frac{\partial g}{\partial \sigma_{uv}}} \tag{2.14}$$

where, g is the plastic potential function; Φ is the yield function; A is the hardening parameter; D is the elastic modulus tensor.

For ideal plastic material, the hardening parameter A in Eq. (2.14) is always zero, and the plastic potential function is the same as the yield function, that is, g and Φ in the equation can be expressed by the function F.

The ideal plastic model can be divided into two types: frictionless constitutive model independent of hydrostatic pressure and frictional constitutive model related to hydrostatic pressure. Friction here refers to the increase of yield limit of material due to hydrostatic pressure. Among them, the representative Mises model without friction and Drucker Prager model with friction are compared. The Mises ideal plasticity model is mainly suitable for metal materials. It does not consider the main constitutive properties of geotechnical materials, such as dilatancy and compressibility, so it is not suitable for geotechnical materials or only suitable for undrained total stress analysis of pure cohesive soil with $\varphi = 0$. Its importance lies in the fact that it is the basis of constructing friction type ideal plastic model and isotropic or non isotropic complex geotechnical constitutive model.

The Drucker-Prager model is established on the basis of generalized Mises yield criterion considering hydrostatic pressure. It is the earliest proposed elastic–plastic model suitable for geotechnical materials. Although it does not reflect the different strength of triaxial tension and compression, hydrostatic pressure can cause the yield of geotechnical materials and the influence of Rhode angle of stress on plastic flow, its greatest advantage of this method is that the influence of hydrostatic pressure on the yield and strength of materials and the dilatancy of geotechnical materials are considered in a simple way, moreover, the model parameters are few and the calculation is simple. Many subsequent isotropic or unequal plastic models of geotechnical materials are established on the basis of modification and expansion. So far, there are many kinds of elasto-plastic models, which are not listed here, but some representative elastic–plastic models are introduced.

Roscoe [5] of Cambridge University proposed the concepts of state boundary surfaces and critical state line. Based on the work hardening theory of plastic mechanics, Roscoe et al. [6] established the elastoplastic model of soil for normal consolidation reshaped clay, called Cambridge model. Burland proposed the modified Cambridge model, and then Roscoe and Burland extended it to the general three-dimensional stress state. Because it is the first soil stress–strain calculation model established by incremental plasticity theory and has achieved certain results, it has become a well-known elastoplastic model of soil.

The central concept of Cambridge model is the concept of state boundary surface. In the *p-q–v* space, the space surface formed by the path of triaxial consolidation drained or undrained test along the normal consolidation curve with the change of consolidation pressure is called Roscoe surface or state boundary interface, but not completely the same as yield surface or loading surface; For normally consolidated or weakly overconsolidated clays, the failure surface is the surface formed by the critical state line and its projection line in the p-q plane. For the strong overconsolidated soil with strain softening property, the failure point is generally at the stress peak point above the critical state line. The plane formed by the intensity peak point in the *p-q–v* space is called the failure surface of this kind of strain-softening material, which is called the Hvorslev surface. After normalization in p-q plane, Roscoe surface, Hvorslev surface and tensionless wall form a complete normalized state boundary line. The model adopts the associated flow law, in which there are three model parameters, *A*, *K* and *M*, which can be measured by conventional triaxial test. The advantage of Cambridge model is that the basic assumptions have certain experimental basis, and the concept is clear. For example, the critical state line, state boundary and elastic wall have clear geometric and physical meanings, and the hydrostatic yield characteristics, compressive stiffness (normal weak overconsolidated clay) and shear dilatancy (strong overconsolidated clay) of geotechnical materials are considered, moreover, the model parameters are easy to obtain. The disadvantage of this model is that the influence of intermediate principal stress is not considered due to the adoption of C-M failure criterion, and there are sharp corners on the failure surface, and the direction of plastic strain increment at the sharp corner is difficult to determine.

Lade [7] and Duncan put forward a lade Duncan model of yield surface, called L-D model for short. The model takes into account the dilatancy and compressibility of the material, the different characteristics of triaxial compression and tensile strength, and the influence of the intermediate principal stress or the Rod angle on yield and failure; The yield surface is smooth and easy to calculate. There are 9 calculation parameters or material parameters in the model, which can be measured by conventional triaxial test. However, the yield surface of the model is a cone with a straight line as the generatrix in the principal stress space, so it can not reflect the characteristics of yield curve changing with hydrostatic pressure, nor can it reflect the yield phenomenon under proportional loading. In order to overcome the shortcoming of single yield surface in L-D model, lade modified the L-D model into lade model with two yield surfaces, which is called L-model for short. Aiming at the deficiency of L-D model, the model is modified in the following aspects.

(1) In order to reflect the yield phenomenon under proportional loading and to overcome the excessive shear dilatancy on the yield surface of linear cone, the yield surface of L-D model is changed to curved conical yield surface fp, and the plastic flow on the yield surface obeys the unrelated flow law.
(2) A spherical cap yield surface fc is added to the cone opening end of the conical surface of the L-D model to consider the yield properties of geotechnical materials under hydrostatic pressure and the shear shrinkage properties of some materials. The plastic flow on fc obeys the orthogonal law.
(3) In order to adapt to the cohesive force or certain tensile strength of overconsolidated clay and other materials, a cohesive force is added to the original normal stress component. The modified L model takes into account the hydrostatic pressure yield characteristics, dilatancy and compressibility of geotechnical materials, the characteristics related to the yield curve and hydrostatic pressure curve, the influence of intermediate principal stress and Rhode angle on Yield and failure, and the uncorrelated flow rule for plastic expansion strain is adopted to avoid excessive dilatancy. The yield surface is smooth, which is conducive to numerical calculation. Although there are 14 parameters in the model, which affect its application in practical engineering, it is more comprehensive in theory and worthy of reference, which is the reason why this book highlights it.

The models introduced above are all elastoplastic model of isotropic hardening. This kind of model is relatively simple, and the differences among the models are mainly reflected in several aspects such as yield criterion, hardening parameter and flow rule. They usually think that when the stress path is in the yield surface, only elastic deformation occurs. However, in fact, when the stress point is in the yield surface, the irreversible plastic deformation may occur with the change of stress state. Especially for the case of periodic loading and reverse unloading, it is more important to consider the possible plastic deformation of the stress path in the yield surface. In order to truly describe the constitutive characteristics of soil under cyclic loading, some researchers have proposed a multiple yield surface model. In the multi yield surface model, the mixed hardening is often used in the yield surface hardening, so the mixed hardening form is summarized first.

The simple form of isotropic hardening can only describe the change of yield surface caused by subsequent loading, and it does not consider Bauschinger effect, initial anisotropy and anisotropy caused by stress of the soil. The reason why it can be widely used in traditional models is that it is more intuitive and clear to a large extent, while the simple movement hardening exaggerates the Bauschinger effect of the soil. The so-called Bauschinger effect means that in the initial deformation stage, the positive yield stress is equal to the reverse yield stress, which is equal to σs, but after strain strengthening, the absolute value of reverse yield stress decreases less than σs. The Bauschinger effect of soil is shown in Fig. 2.3.

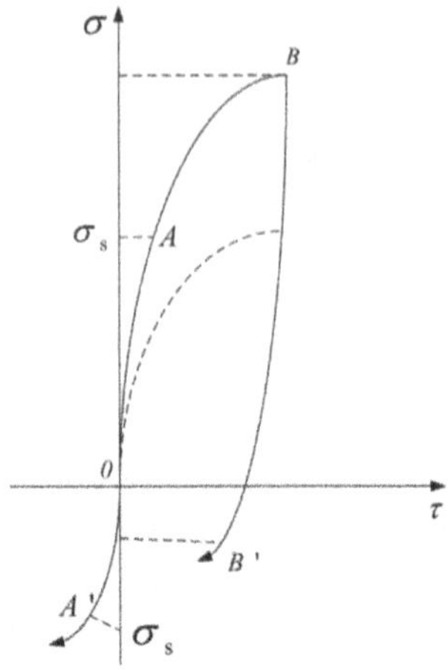

Fig. 2.3 Bauschinger effect of soil

At the same time, the mixed hardening model of isotropic hardening and kinematic hardening is considered. The loading surface of the model can be translated in the stress space, and can be enlarged or reduced with similar shape. When uniaxial compression and then reverse loading, the tensile direction can also be hardened, but the hardening strength is not as large as great of compression, and the degree of weakening is not as strong as that of kinematic hardening. Therefore, the mixed hardening model can best reflect the real characteristics of soil, especially when the soil is subjected to periodic or reverse loading.

Mixed hardening is a combination of isotropic hardening and kinematic hardening. In the traditional elastoplastic constitutive model, the law of isotropic hardening is widely used, while the use of kinematic hardening is less. The following is a brief introduction to the kinematic hardening law which is seldom used at present.

Motion hardening can be roughly divided into the following two types.

(1) Prager's law of hardening

In this hardening law, Prager assumes that the moving stress increment $d\xi ij$ of the central position of the loading surface is related to $d\varepsilon_{ij}^{p}$, namely:

$$\mathrm{d}\xi_{ij} = C\mathrm{d}\varepsilon_{ij}^{p} \tag{2.15}$$

where, C is the proportional coefficient reflecting the material properties. Generally, C is a constant, which can be determined by uniaxial tension and compression experiments. It can be seen from the hardening formula that Prager hardening is linear kinematic hardening, and the direction of $d\xi ij$ and $d\varepsilon_{ij}^{p}$ is the same, which means that the center of loading surface is along the direction of plastic strain increment or the direction of normal outside loading plane.

(2) Ziegler's law of hardening

Prager's law of kinematic hardening can't reflect the special case that the increment of a certain component of σij is not necessarily equal to zero when $d\sigma'_{ij} = 0$. For example, for the conventional triaxial compression experiment, the Prager kinematic hardening law can not be reflected in the case of $d\varepsilon_3^p \neq 0$, $d\varepsilon 3 \neq 0$, while $d\sigma 3 = 0$, $d\sigma 1 \neq 0$. Therefore, Ziegler modified Prager's law of kinematic hardening. In Ziegler's law of hardening, it is assumed that the center of the loading surface moves in the direction of $\sigma ij - \xi ij$, that is:

$$\mathrm{d}\xi_{ij} = d\mu\left(\sigma_{ij} - \xi_{ij}\right) \tag{2.16}$$

where, $d\mu$ is the proportional constant, which is generally a function of generalized plastic strain increasing $d\overline{\gamma} p$. A is a function of the distance between the current stress point and the stress point reached by the next movement or the distance between the two stress points. As Ziegler's law of kinematic hardening can better reflect the hardening characteristics of geotechnical materials, Ziegler's law of kinematic hardening is more commonly used in the non isotropic hardening model of geotechnical materials.

The multiple yield surface model considers that the movement of the yield surface in the stress space is translational and isometric, and it is assumed that there are a cluster of overlapping and disjoint geometrically similar yield surfaces f_0, f_1, f_2, …, fn in the stress space. And each yield surface moves in the stress space in the form of mixed hardening (Fig. 2.4). When the stress moves from the initial point 0 to f, the elastic strain first occurs, and when the stress reaches f_0, the plastic flow begins. Then, f0 moves towards f1 again until they contact. the plastic flow at this time is controlled by the hardening modulus related to f_0. Once f_0 contacts with f_1,f_1 becomes an effective yield surface. This process will continue with further loading.

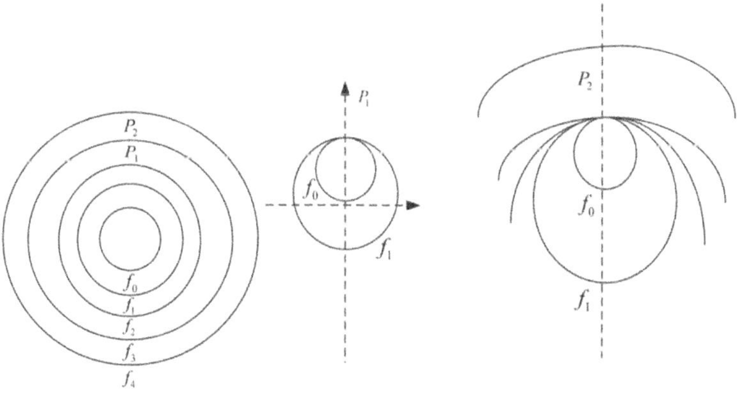

Fig. 2.4 Schematic diagram of the multiple yield surface model and its movement rules in the stress space

Prevost generalized the multi-yield surface model and proposed a generalized multi-yield surface model under drained and undrained conditions. In this kind of model, the overlapped yield surfaces can not only shift, but also shrink and expand in the same direction without crossing each other. Irrelevant flow rules are also used in the drainage model. The equation of multiple yield surfaces in the model is often expressed as:

$$f_i = \frac{3}{2}\left(S_{ij} - \alpha_{ij}^m\right)^2 + C^2\left(P - \beta^m\right)^2 - \left(k^m\right)^2 \neq 0 \tag{2.17}$$

In the generalized multiple yield surface model, since each yield surface represents a certain value of hardening modulus, the position of the yield surface at a certain moment not only reflects the past stress history of materials, but also represents the current size and distribution of plastic modulus field in the stress space. In addition, with the generation and development of plastic strain, the yield surface expands, shrinks and moves in a certain rule. It can not only simulate the anisotropic work hardening characteristics of materials, but also describe the phenomenon of irreversible plastic deformation under cyclic loading, such as unloading nonlinearity, reloading and reverse loading. In a word, although the multiple yield surface model does not break out of the framework of classical plastic mechanics, it can describe some real characteristics of soil which can not be described by classical plastic theory, and has universality and flexibility.

At present, most of the multi yield surface models are mainly based on the characteristics of soil under cyclic and repeated loading and unloading. In fact, the simulation of this characteristic is also the original intention of the original multi yield surface model. Among the existing multi-yield surface models, Morz and Lwan, Pender [8], Dafalias [9–12] and Norris [13] are the most representative. Among them, the models proposed by Morz and Lwan are the earliest. These models systematically give the basic equation of multi yield surface models, and most of the models proposed later are developed on this basis. Although the multi yield surface model has some advantages over the conventional elastoplastic model, it has a disadvantage that can not be ignored, that is, the location, size and plastic modulus of each yield surface must be defined, updated and memorized during calculation, which makes the calculation more complex and requires high memory of the computer, which makes the model difficult to be used in engineering practice. This application inconvenience leads people to find a simpler model, and thus a boundary surface model is proposed. The boundary interface model simplifies the nested yield surfaces in the multi yield surface model. Only one boundary surface and one or two yield surfaces are considered, which inherits the theoretical advantages of the multi yield surface model. In application, the method of tracking the hardening modulus field by the multi yield surface model is replaced by the interpolation equation of plastic modulus, which greatly simplifies the model and enhances its practicability. The boundary surface model was first proposed by Dafalias and Popov [14] and applied to cyclic loading of metal materials. The boundary surface model assumes that there is a boundary surface in the stress space, which limits the movement range of the stress point and

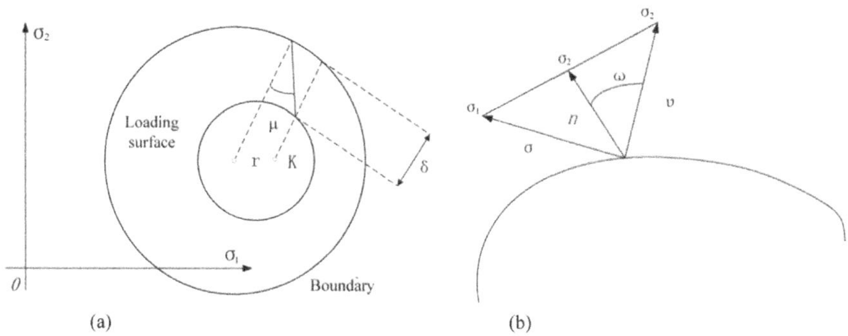

Fig. 2.5 Schematic diagram of boundary surface model and its movement in stress space

yield surface (Fig. 2.5). This surface is mostly elliptical. It contains a yield surface geometrically similar to the boundary surface. In the process of plastic loading, both the boundary surface and the yield surface can expand and contract isotropically and can translate in the stress space. The loading surface can contact and be tangent to the boundary surface, but cannot cross it. The equation of the loading surface is generally defined as:

$$f\left(\sigma_{ij} - a_{ij}, q_n\right) = 0 \tag{2.18}$$

where, aij is the coordinate of the center of the loading surface; qn is the plastic internal variable, such as plastic strain.

The main difference between the boundary model and the multiple yield surface model is that the hardening modulus of the yield surface is a function of the distance between the current stress point on the yield surface and its conjugate point on the boundary surface in the boundary surface model, which is determined by some simple interpolation equation. At present, most of the boundary surface models proposed are two surface models, that is, one boundary surface contains a yield surface. Compared with the multiple yield surface model, the boundary surface model does not need to "memorize" the positions and sizes of multiple yield surfaces, which is relatively simple. However, the interpolation rule used to replace the plastic modulus of the original hardening modulus field has great a priori.

2. **Viscoelastic plastic model**

In order to describe the influence of time on the stress–strain relationship of soil, it is necessary to use a class of time-dependent models to describe the properties of soil, such as viscoelastic model, viscoplastic model, viscoelastic-plastic model and so on. Viscoelasticity is the property of both elasticity and viscosity. The phenomena of creep and stress relaxation are well known and paid special attention to. The characteristic of viscoelasticity is that there are not only stress and strain terms, but also their time derivative terms in the constitutive equation. For linear viscoelastic materials, the general expression of the constitutive equation is:

$$a0\alpha + a1\dot{\sigma} + \cdots + am\sigma^{(m)} = b0\varepsilon + b1\dot{\varepsilon} + \cdots + bn\varepsilon^{(n)} \tag{2.19}$$

where, ai and bi are parameters related to material properties. The commonly used and simple viscoelastic models are Maxwell model, Kelvin model and three element viscoelastic model. Maxwell model is also called relaxation model. It consists of a linear spring and a Newtonian adhesive pot in series. The stress acting on the two elements is the same, and the total strain is the sum of the two elements. The constitutive model of the material described in this model is:

$$\frac{\mathrm{d}\varepsilon}{\mathrm{d}t} = \frac{1}{E}\frac{d\sigma}{dt} + \frac{\sigma}{\eta} \tag{2.20}$$

The Maxwell model can only describe the relaxation characteristics of materials, but not the creep characteristics of materials. The Maxwell's body has no aftereffect and flows at a fixed speed when the shear stress is constant, which is equivalent to Newtonian ideal viscous liquid. Kelvin body, also known as creep viscoelastic body, is composed of a linear spring and a Newtonian adhesive pot in parallel. Under the condition of parallel connection, the strains of the two elements are the same, and the total stress is the sum of the stresses of the two elements. The constitutive equation of the Kelvin model is $\sigma = E\varepsilon + \eta dS/dt$, Kelvin body can be regarded as an elastic solid with viscous property and has aftereffect property, but it has no relaxation property. When the shear strain is constant, its properties are equivalent to that of elastic solid. Kelvin model can also be used to represent the damping oscillation caused by internal friction, that is, the energy loss caused by damping. The three-element model can more comprehensively reflect the characteristics of linear viscoelastic solids. It has different composition methods. Among them, the three-element model composed of linear spring and Kelvin model in series is called L-body model for short, and the three-element model composed of sticky pot and Maxwell model in parallel is called j-body model for short. The constitutive equation can be uniformly expressed as:

$$\sigma + p1\dot{\sigma} = q1\dot{\varepsilon} + q2\ddot{\varepsilon} \tag{2.21}$$

For L-body model:

$$p_1 = (\eta_1 + \eta_2)/E_1, q_1 = 2\eta_2, q_2 = 2\eta_1\eta_2/E_1$$

For J-body model:

$$p_1 = \eta_1/E_1, q_1 = (\eta_1 + \eta_2), q_2 = 2\eta_1\eta_2/E_1$$

In addition to the above models, there are more complicated viscoelastic plastic models for the comprehensive simulation of soil properties, which will not be discussed here.

3. **Plastic Endochronic Theory**

In the classical plastic theory, it is always assumed that there is a yield surface or a loading surface corresponding to the hardening or softening process. However, the experiments of geotechnical materials show that there is no obvious yield point in theory when the soil is compressed or sheared, and the residual deformation often occurs from the beginning of loading. Therefore, in this sense, the simplified yield surface theory is often different from the real situation of soil deformation. Valanis [15] proposed the plastic endochronic theory (plastic endochronic theory) which also used incremental theory, but there is no yield surface. It was first proposed for metal materials, and then Bazant and Valanis popularized it to engineering materials such as geotechnical engineering and concrete. The plastic endochronic theory is equivalent to the mixed theory of classical linear viscoelastic theory and incremental plastic theory. Its basic feature is the introduction of the concept of internal time and internal time measurement reflecting the cumulative nonlinear strain of the material. and the change of constitutive relation caused by the change of endochronic system replaces the yield criterion and hardening law of incremental plasticity theory without the concept of yield surface. Endochronic theory is a relatively new theoretical system, which is still improving and developing.

With the development of soil constitutive model, the research of soil constitutive model under simple loading condition has become more and more mature, while the research on soil constitutive model under complex loading conditions is still under development. At present, the research of soil constitutive model tends to consider the complex stress path and the constitutive relationship under complex loading condition, and most of the models have strong pertinence, rather than a general and versatile constitutive model of soil.

2.3 Constitutive Model of Expansive Soil

Soil mechanics is a discipline that combines engineering mechanics with soil properties. This broad expression can cover all kinds of soils. But the classical soil mechanics focuses on silt, clay and sand in its development process. There are many kinds of soils encountered in engineering practice, and their properties do not conform to the principles and concepts of classical soil mechanics. Expansive soil is one of the representative soils. Expansive soils, as a verification of special clays, do not conform to the deformation theory of general clays. Expansive soil undergoes water migration and transformation with changes in climate, geological environment, and engineering construction activities. The properties of expansive soils change with the change of water, showing different characteristics. The variability of shear strength of expansive soils is mostly related to the action of water. Expansive soil absorbs water and causes anisotropic expansion, which may cause deviation stress. The uneven expansion caused by different water absorption distribution conditions may also cause the destruction of expansive soil.

2.3.1 Strength Theory of Expansive Soil

In the 1930s, Terzaghi [16] proposed the concept of effective stress. The stress state variables that control the properties of saturated soil are described as follows: The stress at any point in the soil can be calculated from the total principal stress $\sigma_1, \sigma_2, \sigma_3$ acting on that point, the stress in water is u_w. The difference between the two is the effective stress:

$$\sigma' = \sigma - u_w \tag{2.22}$$

After Tasighi proposed the concept of effective stress, some effective stress equation for saturated soils have been put forward successively by scholars, of which Bishop's expression has the greatest influence [17]:

$$\sigma' = (\sigma - u_a) + \chi(u_a - u_w) \tag{2.23}$$

where, u_a is pore pressure and χ is a parameter related to soil saturation. For saturated soils, $\chi = 1$; for dry soils, $\chi = 0$. The relationship between χ and saturation is determined experimentally.

Since then, Aitchison [18, 19] proposed the following effective stress equation:

$$\sigma' = \sigma + \psi p'' \tag{2.24}$$

where, p'' is the difference in pore water pressure; ψ is a parameter ranging from 0 to 1.

Richard adds a component of solute suction to the effective force equation:

$$\sigma' = \sigma - u_a + \chi_m(h_m + u_a) + \chi_s(h_s + u_a) \tag{2.25}$$

where, χ_m is the effective stress parameter considering the matrix suction; χ_s is the effective stress parameter considering the solute suction; h_m is the matrix suction; h_s is the solute suction. This equation is rarely mentioned later because parameter χ is too complex.

The above research shows that scholars have made great efforts to establish a single-value effective stress equation for expansive soils. But the experimental results show that the suggested effective stress equation is not a single value, but is related to the stress path. The parameter χ of effective stress is difficult to determine. These equation have little effect in practice. The result of the parameter χ measured by strain is inconsistent with the result measured by shear strain. After reassessment of the various effective stress equation, many researchers prefer to use two independent stress state variables $\sigma - u_a$ and $u_a - u_w$ to describe the mechanical properties of unsaturated soils.

Fredlund [20, 21] and Morgenstern proposed an expansive soil stress analysis based on continuous multiphase media mechanics, the equation is as follows:

$$\tau_f = c' + (\sigma - u_a)\tan\varphi' + (u_a - u_w)\tan\varphi^b \tag{2.26}$$

where, c' is the effective cohesion; φ' is the effective internal friction angle; φ^b is the internal friction angle whose strength changes with suction.

Shen Zhujiang obtained the following equation by modifying χ:

$$\tau_f = c' + (\sigma - u_a)\tan\varphi' + \frac{u_a - u_w}{1 + d(u_a - u_w)}\tan\varphi'. \tag{2.27}$$

2.3.2 Deformation Theory of Expansive Soil

1. **One-dimensional deformation theory**

The bulk deformation and strength characteristics of expansive soils are not simply controlled by σ, u_a and u_w, but it is controlled by u_a and $u_a - u_w$. Volumetric variation of soil reflects strains of soil particles, water and air at the same time. Fredlund [22, 23] obtained according to the one-dimensional compression experiment.

$$e = e_0 - c_t \lg\frac{(\sigma - u_a)_f}{(\sigma - u_a)_0} - c_m \lg\frac{(u_a - u_w)_f}{(u_a - u_w)_0} \tag{2.28}$$

$$w = w_0 - d_t \lg\frac{(\sigma - u_a)_f}{(\sigma - u_a)_0} - d_m \lg\frac{(u_a - u_w)_f}{(u_a - u_w)_0} \tag{2.29}$$

where, e is the void ratio; w is the water content; 0 and f are the initial state and final state of the test respectively; c is the compressibility coefficient; d is the water content index.

The deformation of expansive soil can be divided into the deformation of soil particles, water and gas, and the corresponding incremental form can be obtained:

$$d\varepsilon = m_1^s d(\sigma - u_a) + m_2^s d(u_a - u_w) \tag{2.30}$$

$$d\theta_w = m_1^w d(\sigma - u_a) + m_2^w d(u_a - u_w) \tag{2.31}$$

$$d\theta_a = m_1^a d(\sigma - u_a) + m_2^a d(u_a - u_w) \tag{2.32}$$

Loret derives the following formulas from compression tests of pore pressure, pore water pressure and water content:

$$d\theta_a = m_1^a d(\sigma - u_a) + m_2^a d(u_a - u_w) \tag{2.33}$$

$$e = a_2 + b_2(\sigma - u_a) + c_2 \lg(u_a - u_w) + d_2(\sigma - u_a)\lg(u_a - u_w) \tag{2.34}$$

$$S = a_3 - th|b_3(u_a - u_w)[c_3 + d_3(\sigma - u_a)]| \tag{2.35}$$

$$S = a_4 - |1 - \exp[-b_4(u_a - u_w)]|[c_4 + d_4(\sigma - u_a)] \tag{2.36}$$

where, e is void ratio; S is saturation.

2. **Three-dimensional deformation theory**

(1) **Elastic–plastic model**

The elasto-plastic constitutive models of expansive soils are developed from the elasto-plastic constitutive models of saturated soils.

Based on the Cambridge model, Alonso [24] adds a suction parameter to the stress–strain space to form a generalized yield surface:

$$F_1 = q^2 - M^2(p + p_s)(p_0 - p) = 0 \tag{2.37}$$

$$F_2 = u_s - u_{s0} = 0 \tag{2.38}$$

The plastic deformation is taken as the hardening parameter. After derivation, the incremental form of volume strain ε_v and generalized shear strain ε_s can be obtained:

$$d\varepsilon_v = \frac{k}{v}\frac{dp}{p} + \frac{\kappa_s}{v}\frac{du_s}{u_s + p_{at}} + \mu_1 + \mu_2 \tag{2.39}$$

$$d\varepsilon_s = \frac{dp}{3G_E} + \mu_1 \frac{2qa}{M^2(2p + p_s - p_0)} \tag{2.40}$$

where, $v = 1 + e$, e is the void ratio; G_E is the elastic shear modulus.

$$\mu_1 = \frac{\frac{\partial F_1}{\partial p}\mathrm{d}p + \frac{\partial F_1}{\partial q}\mathrm{d}p + \frac{\partial F_1}{\partial u_s}\mathrm{d}u_s}{\frac{\partial F_1}{\partial p(0)}\frac{p(0)}{\partial \varepsilon_v^p}} \tag{2.41}$$

$$\mu_2 = \frac{-\frac{\partial F_2}{\partial u_s}\mathrm{d}u_s}{\frac{\partial F_2}{\partial u_{s0}}\frac{\partial u_{s0}}{\partial \varepsilon_v^p}} \tag{2.42}$$

The advantage of the elastic–plastic constitutive model is that the parameter setting is easy, but the disadvantage is poor adaptability. It is difficult to describe the properties of unsaturated soils and the equation is based on the test results of disturbed soils and cannot reflect the strain softening. According to the hardening law, the volume shrinks when hardening and expands when softening. The softening of expansive soils is accompanied by volume shrink age, it directly violates the hardening law. Therefore, the elastoplastic constitutive model is only applicable to soils without obvious softening characteristics or super-consolidated soil with softening properties.

(2) **Generalized suction model**

Shen Zhujiang believes that the role of suction is to increase the anti-sliding resistance between soil particles, and define all factors that increase the anti-skid resistance as generalized suction. The generalized suction is mainly composed of reduced suction and cemented suction. The essence of reducing suction is matrix suction, and the cementation suction is the reflection of structural strength, Shen Zhujiang assumed that the law of loss of reduced suction and cemented suction is expressed as:

$$u_{\mathrm{se}} = u_{\mathrm{se}0}[(w_s - w)/(w_s - w_0)]^n \tag{2.43}$$

$$u_{\mathrm{sb}} = u_{\mathrm{sb}0}\left[1 + \cos\pi\left(\frac{q - q_0}{q_{\max} - q_0}\right)\right] \tag{2.44}$$

where, u_{se0} is the reduced suction of the undisturbed soil; u_{sb0} is the suction of the undisturbed soil structure; q is the generalized sliding force.

$$q = (a\sigma_{\mathrm{m}} + b)/\left(\sigma_{\mathrm{m}} + u_s'\right) \tag{2.45}$$

where, σ_m is the normal stress; σ_s is the tangential stress and u_s is the general suction. The generalized suction model divides the strain into two parts, one part is caused by the effective stress, and the other part is caused by the loss of suction, the deformed expression is expressed as:

$$\Delta\varepsilon = C_{EP}\Delta\sigma' + c_{\mathrm{b}}(e - e_s)\Delta u_s' I + c_s \Delta u_s' \overline{S}/\sigma_s \tag{2.46}$$

where, C_{EP} is the elastoplastic flexibility matrix; e is the actual void ratio; e_s is the saturated soil void ratio. The advantage of this model is that it is more adaptable, but the disadvantage is that the law of suction loss is difficult to determine.

(3) **Damage mechanics model**

Shen Zhujiang proposed a damage mechanics model for expansive soil, and used it to describe the shear softening characteristics of structured clays and loess with high void ratios. The model assumes that the undisturbed soil-like linear elastic body becomes disturbed soil after being completely damaged. This change process can be expressed as:

$$\sigma = (1 - \omega)D_{\mathrm{i}}\varepsilon + \omega \mathrm{D}_{\mathrm{s}}\varepsilon \quad (2.47)$$

where, D_i is the stiffness matrix of the undisturbed soil; D_s is the stiffness matrix of the disturbed soil; ω is the damage ratio, i.e. the proportion of disturbed soil. Shen Zhujiang believes that this model can reflect the phenomena of strain softening and water immersion softening, but whether this model can be popularized and applied and the disturbed soil is regarded as complete damage remains to be verified.

It can be seen from the research history of the constitutive model of the expansive soils mentioned above. Since Bishop [17] proposed the effective stress constitutive model, subsequent studies have not deviated from its model. Only with the deepening and promotion of research, the definition of model parameters becomes more and more elaborate and complex. Researchers spend a lot of time guessing the mathematical expression of the constitutive model of expansive soil. But no matter which model is widely recognized and cited, this situation has been plagued by the development of expansive soil mechanics. Therefore, it is an urgent task to adopt new modeling methods in geotechnical forces.

2.4 Influence of Stress Path on Constitutive Relationship of Rock and Soil

Stress path refers to the process of stress, occurrence, development and change of soil mass, which is actually a track line of stress change. It is easy to misunderstand various mechanical indexes obtained by conventional tests. First, the stress state of soil mass in actual engineering is quite different from that stipulated by laboratory test. Second, the nature of soil is very complex, its characteristics are not only related to the magnitude of the force, but more importantly, it is related to the way and method of force application, the sequence and a series of changes. Therefore, the properties of soil can be reflected by unconventional tests. In order to make the problems studied as close as possible to the actual conditions, both indoor and field tests should simulate the actual stress process of soil mass as possible. The researcher masters the analysis method of stress path, which is especially helpful for experimental design according to actual situation and obtains corresponding indexes for theoretical analysis.

The stress path of the triaxial test can vary according to the actual situation. It is much more flexible than unidirectional compression test and direct shear test soil samples under the defined path.

Conventional triaxial tests are performed by loading the specimen to shear failure under axisymmetric conditions. The loading method is divided into triaxial compression test and triaxial tension test. The combination of stresses applied by these two tests is as follows.

Compression test:

(1) Fixing σ_3, increasing σ_1, and increasing the sum of principal stress;
(2) Fixing σ_1, reducing σ_3, and reducing the sum of principal stress;
(3) The sum of principal stress remains unchanged, increase σ_1, reducing σ_3;

Tensile test:

(1) Fixing σ_3, reducing σ_1, and reducing the sum of principal stress;
(2) Fixing σ_1, increasing σ_3, and increasing the sum of principal stress;
(3) The sum of principal stress remains unchanged, reducing σ_1,increasing σ_3;

The combination of several stress paths is shown in Fig. 2.6. A brief description of these stress paths is given below.

1. **Equal Proportional Loading Path Test**

The test starts from the initial hydrostatic pressure state ($\sigma_1^0 = \sigma_2^0 = \sigma_3^0 = \sigma_c$). Then the soil element undergoes stress increment (or change) in the three principal stress directions $\Delta\sigma_1$, $\Delta\sigma_2$, $\Delta\sigma_3$. Let $\sigma_1 : \sigma_2 : \sigma_3 = a_1 : a_2 : a_3$, Where a_1 and a_2 are parameters to determine the NP direction of the straight path in Fig. 2.6. Consider all stress paths described in Fig. 2.6 (load compression (CTC), loaded tension (CTE), average principal stress constant compression (TC), average principal stress constant tension (TE), hydrostatic pressure (HC))as a special example of a conventional linear stress path, the different values of parameters a_1 and a_2 are used for each test. For example, the HC path corresponds to $a_1 = a_2 = 1$, and the CTC path corresponds to $a_1 = a_2 = 0$.

2. **Hydrostatic pressure test**

The stress path corresponding to this test remains in the direction of the hydrostatic pressure axis ($\sigma_1 = \sigma_2 = \sigma_3$), as shown in Fig. 2.6. In this stress path, the changes in stress invariants I_1 (or p) and J (or q) are expressed by Eq. (2.48) as:

$$\Delta I_1 = 3\Delta\sigma_{oct} = 3\sigma_c,\ \Delta J_2 = \Delta\tau_{oct} = 0 \tag{2.48}$$

where, $\Delta\sigma_c$ is the increment of hydrostatic pressure, which mainly provides the body strain information of soil.

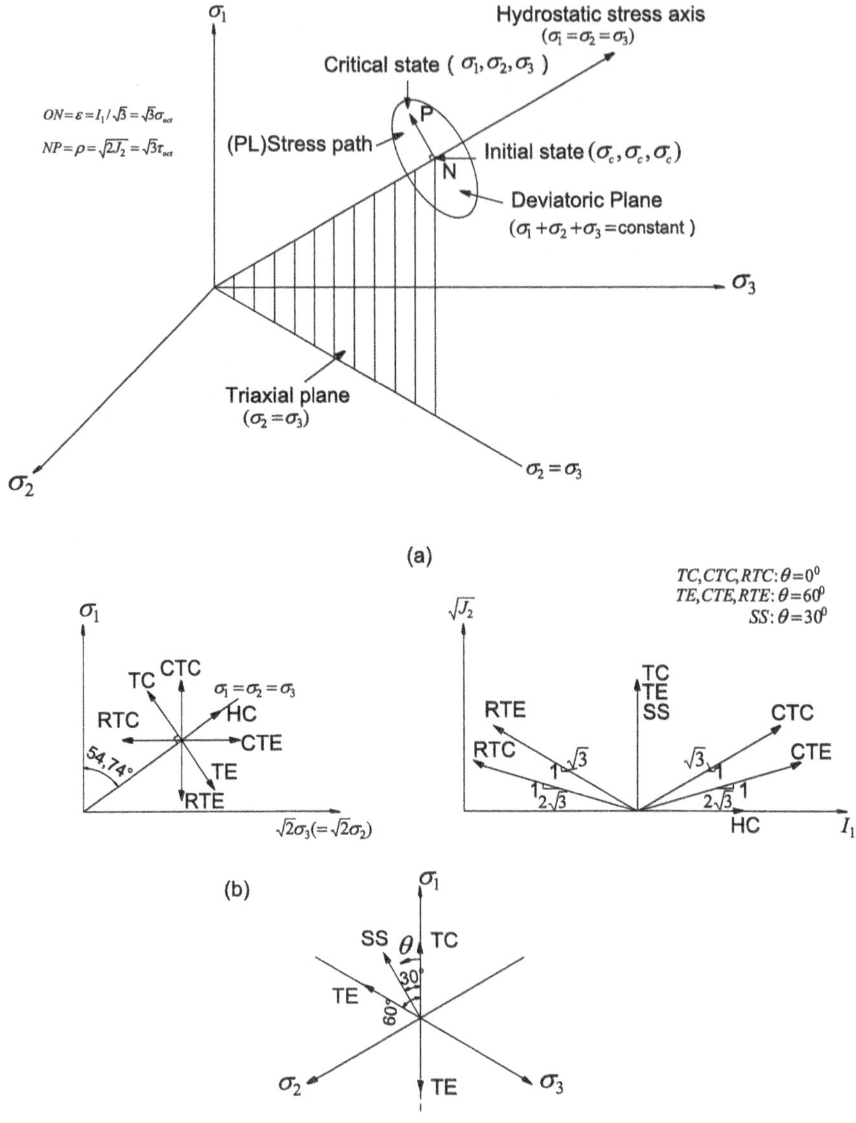

Fig. 2.6 Combination of several stress paths

3. Conventional triaxial pressure test

This test method is most commonly used, the two principal stresses of the specimen are $\sigma_2 = \sigma_3 = \sigma_c$. The stress path corresponding to the increase in the third principal stress σ_1 corresponds to the CTC as described in the triaxial plane and meridian plane of Fig. 2.6 respectively. In this case, σ_1 is the maximum principal stress, and σ_2 and σ_3 are the middle and minimum principal stresses, respectively. The average principal normal stress and generalized shear stress increments $\Delta\sigma_{oct}$ and $\Delta\tau_{oct}$ are expressed as follows:

$$\Delta\sigma_{\text{oct}} = \frac{\Delta I_1}{3} = \frac{\Delta\sigma_1}{3}, \Delta\tau_{\text{oct}} = \left(\frac{2}{3}\Delta J_2\right)^{1/2} = \frac{\sqrt{2}}{3}\Delta\sigma_1 \tag{2.49}$$

where, $\Delta\sigma_1$ is the change value of the large principal stress σ_1. Therefore, the slope of the path, $\sqrt{\Delta J_2}/J_1 = 1/\sqrt{3}$ represents the CTC in Fig. 2.6.

4. **Conventional Triaxial Tensile Test**

The test is performed when σ_1 remains unchanged, and the increase of equivalent values of σ_2 and σ_3 causes σ_1 to become the minimum principal stress. In this case, $\Delta\sigma_{oct}$ and $\Delta\tau_{oct}$ are respectively represented as:

$$\Delta\sigma_{\text{oct}} = \frac{\Delta I_1}{3} = \frac{2\Delta\sigma_2}{3}, \Delta\sigma_{\text{oct}} = \left(\frac{2}{3}\Delta J_2\right)^{1/2} = \frac{\sqrt{2}}{3}\Delta\sigma_2 \tag{2.50}$$

where, $\Delta\sigma_2$ is the change value of the large principal stress σ_2. Therefore, $\sqrt{\Delta J_2}/\Delta I_1 = \frac{1}{2\sqrt{3}}$ represents the slope of the CTE path in Fig. 2.6.

In addition to the above four more common stress paths, there are also the following situations: Reduced triaxial compression test (RTC), reduced triaxial tensile test (RTE), triaxial compression and triaxial tensile test, simple shear test (SS) and various other triaxial tests. These are shown in Fig. 2.6b. In this book, triaxial tests on sandy soils are carried out using a stress path of 1–3. Due to time and test conditions, other stress path tests will not be discussed.

References

1. Zhujiang S (1996) Generalized suction and unified deformation theory of unsaturated soils. J Geotech Eng 18(2):49–55
2. Zhujiang S (2000) Masonry model of structural clay. Geotech Mech 21(1):1–4
3. Maohong Y, Liu F (1990) A new universal strength theory. Acta civil Eng Sinica 23(1):34–40
4. Maohong Y, Songyan Y (1997) Twin shear unified elastoplastic constitutive model and its engineering application. Chin J Geotech Eng 19(6):2–10
5. Roscoe KH (1968) On the generalised stress-strain behaviour of wet clay. Eng Plast: 535–609
6. Roscoe KH, Schofield AN, Thurairajah A (2015) Yielding of clays in states wetter than critical. Géotechnique 13(3):211–240
7. Lade PV, Inel S (1997) Rotational kinematic hardening model for sand. Part 1 concept of rotating yield and plastic potential surfaces. Comput Geotech 21(3):183–216
8. Pender MJ (1978) A model for the behavior of overconsolidated soil. Geotechnique 28(1):25–51

9. Dafalias YF (1979) A model of soil behavior under monotonic and cyclic loading conditions. In: 5th international conference on structure mechanics in reactor technology, London
10. Dafalias YF, Manzari MT, Papadimitriou AG (2002) A simple anisotropic clay plasticity model. Mech Res Commun 29(4):241–245
11. Dafalias YF, Manzari MT (2004) Simple plasticity sand model accounting for fabric change effects. J Eng Mech 130(6):622–634
12. Dafalias Y, Taiebat M (2013) An atomy of rotational hardening in clay plasticity. Geotechnique 63(16):1406–1418
13. Norris VA (1982) Numerical modeling of soil response to cyclic loading "stress-reversal surfaces". In: International symposium numerical models in geomechanics, Zurich, pp 38–49
14. Dafalias YF, Popov EP (1975) A model of non-linearly hardening materials for complex loadings. Acta Mech 21(3):173–192
15. Valanis KC (1980) Fundamental consequences of a new intrinsic time measure plasticity, as limit of the endochronic theory. Arch Mech 32:67–91
16. Terzaghi KV (1936) The shear resistance of saturated soil. In: 1st international conference on soil mechanics, pp 54–56
17. Bishop W (1959) The principle of effective stress. Lecture delivered in Oslo. Geotechnique 106(39):859–863
18. Aitchison GD (1961) Relationship of moisture and effective stress functions in unsaturated soils. In: Pressure and suction in soils conference, London, pp 47–52
19. Aitchison GD (1965) Statement of the review panel: engineering concepts of moisture equilibria and moisture changes in soils beneath covered area. In: Proceedings moisture equilibria and moisture changes in soils beneath covered areas, Butterworths, pp 7–21
20. Fredlund DG, Rahardjo H (1993) Soil mechanics for unsaturated soils. Wiley, NewYork
21. Fredlund DG, Rahardjo H (1997) Mechanics of unsaturated soils. Trans. Chen Zhongyi, et al. Architectural Industry Press, Beijing
22. Fredlund DG, Hung VQ (1997) Prediction of volume change in an expansive soil as a result of vegetation and environmental changes. Geotech Spec Publ 115:24–43
23. Fredlund DG (1997) An introduction to unsaturated soil mechanics. Proc Unsaturated Soil Eng Pract Geotech Spec Publ 68:1–37
24. Alonso EE, Gen A (1990) A constitutive model for partially saturated soils. Geotechnique 40(3):220–229

Chapter 3
Numerical Modeling Methods

In Chap. 2, the establishment process of the rock-soil constitutive model is discussed in detail. It can be seen from it, whether these models are elastic, plastic, elastoplastic, or other theoretical forms, these are all regarded as empirical. Although it is solvable in some practical engineering situations, it is difficult to obtain analytical solutions due to the many complexities in the natural state of geological bodies. For inhomogeneous media, non-linear material properties, in-situ stress conditions, spatial and temporal variations of material properties, and the arbitrariness of geometric shapes, discontinuities, and some other complex problems caused by geological features, this method cannot produce real solutions.

3.1 Inverse Problem Theory and Application of Geotechnical Constitutive Relation

In the process of human being's understanding of the world, the occurrence and development of something will cause the occurrence and change of something else, named causality. If the cause of a thing is known to infer the result of the thing, it is called a positive problem; otherwise, it is called an inverse problem to find the cause from the result.

From the point of view of thinking mode, traditional geotechnical mechanical analysis method is a positive thinking. That is, proceeding from the necessity of things, models are established based on experiments to deal with the constitutive relation, solve under certain limited conditions. This is reflected in the study of parameters, which is sampling-design test-measurement-result analysis. Reflected in the study of the model is based on the existing axioms and theorems, plus the assumptions under certain conditions, to obtain the results through deduction. Positive problem analysis requires sufficient and accurate data. Displacement back analysis method developed in 1970s is an inverse problem of reverse thinking. It is based on the measured displacement value to obtain the rock mechanics parameters and initial

Q. Ren and J. Zhou, *Numerical Modeling of Soil Constitutive Relationship*,
https://doi.org/10.1007/978-981-16-3231-0_3

ground stress through inversion. It opens up a new way to study rock mechanical parameters and initial ground stress.

3.1.1 General Description of Inverse Problems

Inverse problems are generally divided into two main categories: One is medium inverse problem, inversion of physical and mechanical parameters of medium. It corresponds to coefficient inversion in differential equations. For example, constitutive model problem and stress wave detection of pile foundation are media inverse problems. The second is the inverse problem of force source, which corresponds to the inversion of inhomogeneous terms in differential equations.

3.1.2 The Inverse Problem of Constitutive Relations

In fact, the process of establishing geotechnical constitutive model is to retrieve the inherent laws of geotechnical media from their mechanical behavior in limited tests. So it is an inverse question. The establishment of geotechnical constitutive model can be roughly divided into four steps.

(1) Determine the basic framework of the model

First, the basic framework of the constitutive model is selected, namely, the type of the constitutive model. According to the characteristics and accuracy requirements of different rock-soil constitutive relations, different types of models can be selected. Such as elastic model, nonlinear model, elastoplastic model, etc.

(2) Parameterization of the model

Constitutive model is a macro description of the phenomena of the constitutive relationship. Based on the established model framework, choose a set of minimum parameters that fully characterize the constitutive relationship, called model parameters.

(3) Reverse simulation

Some experiments are designed to invert the actual values of model parameters by using the measured results of some observable parameters.

(4) Forward simulation

The established model predicts the measurement results of some observable parameters. This is validation of the model.

It should be pointed out here that these four steps are not completed in isolation and one-time, and there is feedback and interaction between each other. The choice of model parameters is not unique.

3.2 Principle of Interaction Between Plastic Body Strain and Plastic Shear Strain

Geotechnical materials are natural geological materials and mainly consist of some mineral particles. There are many cracks and pores between the particles, so the cohesion between the particles is very low. Its shear resistance is mainly dependent on internal friction related to normal stress, so it is also called friction material. It is precisely because of their low ability to resist deformation. Therefore, they are extremely susceptible to external loads and the surrounding environment. It exhibits many unique deformation characteristics and is one of the most complex materials in solid mechanics.

Three basic mechanical characteristics are displayed during the deformation of rock and soil: Compressibility, dilatancy and stress path dependence. The first two characteristics are absent in metal deformation. In metals, the stress path correlation is only shown by the difference between loading, unloading and reloading. For compressibility, as the external confining pressure increases, the shear resistance of the soil increases. When a hydrostatic pressure is applied to the soil, as the soil is filled with many voids. It's these voids that are compressed first, as voids decrease, the contact area between particles increases, thus increasing the internal friction. As a result, the shear resistance increases.

For compressibility, the shear capacity of soil increases with the increase of external confining pressure. When a hydrostatic pressure is applied to the soil, since the soil is filled with many voids, the first one to be compressed is these voids. The voids decrease, the contact area between particles increases, and the internal friction increases. In this way, the shear resistance increases.

For dilatancy, it includes two aspects: Shear shrinkage and dilatancy. During shear deformation, the particles in the soil can be rearranged and distributed. Under certain conditions, the volume shrinks. In the opposite case, the bite or crush of particles in the soil, the expansion of primary fissures or the formation of new fissures in rocks, which results in volume expansion under certain conditions. The certain conditions mentioned here are related to two factors, namely the degree of soil compaction and the external surrounding pressure. The occurrence of shear expansion and contraction depends on the combination of these two factors.

For the stress path correlation, the constitutive relationship of the soil is related to the stress path. Based on experiments and theoretical analysis, it has been proved that. The correlation of the stress path is essentially the comprehensive performance of compressibility and dilatancy. That is, the stress path dependence includes both compressibility and dilatancy.

In order to theoretically explain the mechanism of the above-mentioned soil deformation characteristics, at present, the cross action of average normal stress p and generalized shear stress q is generally used to explain the compression and dilatancy. That is, the direct effect of p on shear strain and the direct effect of q on volume strain. However, this cross-effect of p and q cannot explain the non-compressibility of metal deformation. Since p can directly act on shear strain, compressibility can also occur in metals. It should be noted that due to the dense metal, there is little plastic volume deformation at 30,000 atmospheres.[1] According to the local action principle in Noll's constitutive equation, the generation and transfer of stress in solid deformation are realized by the deformation of the object. It can be understood that the effect of p on shear strain is not direct. But it is achieved through body strain. The plastic volume strain in the metal is approximately zero, so p cannot exert an influence on the shear strain through it.

It is concluded that: The interaction of p and q is essentially the interaction between plastic shear strain and plastic body strain. The mode of such interaction and its effect on shear resistance are clearly shown in Fig. 3.1.

As can be seen from Fig. 3.1, the action mode of plastic body strain on plastic shear strain is completely different from the action mode of plastic shear strain on plastic body strain. The role of the former is not direct, but affects the shear strain by changing the shear resistance. Shear strain affects volume strain through dilatancy and shrinkage. However, in all cases, the change in shear strength is due to the action of plastic body strain. Therefore, it can be concluded that the change of plastic body strain controls the rise and fall of shear resistance.

From the above analysis, the following conclusions can be drawn: The interaction between plastic volume strain and plastic shear strain is the source of the three basic deformation characteristics of rock and soil. At the same time, it is the main source of complexity and diversity of geotechnical mechanical response.

The goal of establishing geotechnical constitutive model is to describe the deformation characteristics of geotechnical soils comprehensively and accurately. To achieve this goal, according to the interaction principle between plastic body strain and plastic shear strain. The constitutive model should adequately reflect this interaction.

Fig. 3.1 ε_v^p-ε_s^p interaction

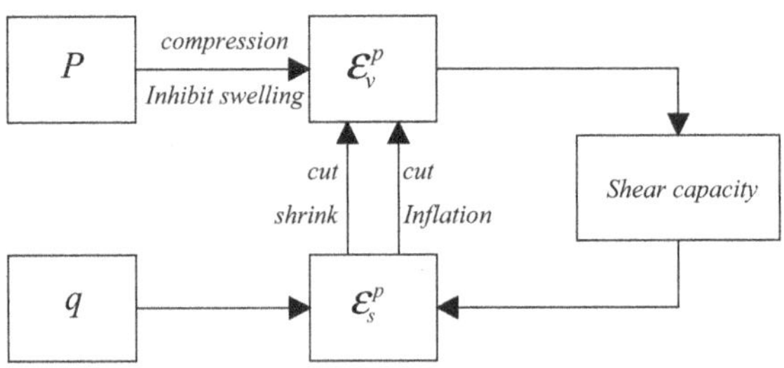

[1] 1 atmospheric pressure = 0.1013 MPa.

3.3 Basic Theory of Modeling

At present, the traditional modeling method of elastic–plastic constitutive relation of rock and soil is based on plastic potential theory. Some experimental results for certain types of soil, the analytical expression of the plastic potential function is found and the model parameters are determined by experiments. In the theory of metal plasticity, since plastic deformation of metals is purely shear deformation. Therefore, it is easier to find a suitable analytical expression of the plastic potential function, such as the Mises yield function. The plastic deformation of rock and soil is quite complex. According to the principle of interaction between plastic body strain and plastic shear strain. The plastic deformation of rock and soil exhibits many unique properties, the most important of these are compressibility, dilatancy and stress path correlation. According to traditional modeling methods, it is obviously very difficult to reflect these characteristics, especially the correlation of stress paths.

Different from traditional modeling methods, numerical modeling is the opposite. It is not the first step to find the analytical expression of the plastic potential function, but to invert the constitutive law directly from the test results. The constitutive equation thus obtained can accurately reflect the basic characteristics of geotechnical mechanical response. In fact, this inversion process is a mapping problem. That is, the "data space" consisting of a large amount of data obtained from experiments is mapped to the corresponding "model space". According to the inverse problem theory, four basic steps of geotechnical numerical modeling are proposed. The general expression of the constitutive relation derived above is taken as the basic constitutive frame of numerical modeling, which lays a theoretical foundation for numerical modeling.

3.3.1 *The Basic Framework of Numerical Modeling Methods*

The stress–strain relationship of soils is expressed by the following two functional functions:

$$\varepsilon_{\mathrm{v}} = F_1(p, q, p(q)) \tag{3.1}$$

$$\overline{\varepsilon} = F_2(p, q, p(q)) \tag{3.2}$$

This stress–strain relationship reflects the coupling relationship between volume strain and shear strain and the cross effect of p and q. It not only considers the dilatancy and shrinkage of the soil, but also reflects the influence of the stress path on the constitutive relationship. As the model parameters in this book are obtained by experimental simulation based on the actual stress path of the soil mass, therefore, the influence of the stress path on the constitutive relationship has been considered, so the Eqs. (3.1) and (3.2) can be simplified as:

$$\varepsilon_v = f_1(p, q) \tag{3.3}$$

$$\overline{\varepsilon} = f_2(p, q) \tag{3.4}$$

In the form of increment:

$$d\varepsilon_v = \frac{\partial f_1}{\partial p}dp + \frac{\partial f_1}{\partial q}dq \tag{3.5}$$

$$d\overline{\varepsilon} = \frac{\partial f_2}{\partial p}dp + \frac{\partial f_2}{\partial q}dq \tag{3.6}$$

Assuming:

$$A = \frac{\partial f_1}{\partial p}, B = \frac{\partial f_1}{\partial q}, C = \frac{\partial f_2}{\partial p}, D = \frac{\partial f_2}{\partial q}$$

Equations (3.5) and (3.6) can be changed into:

$$\begin{Bmatrix} dp \\ dq \end{Bmatrix} = \frac{1}{AD - BC}\begin{bmatrix} D & -B \\ -C & A \end{bmatrix}\begin{Bmatrix} d\varepsilon_v^p \\ d\varepsilon_s^p \end{Bmatrix} \tag{3.7}$$

In order to establish the constitutive equation expressed in Eq. (3.7), the key problem is to determine two unknown functions f_1, f_2 and their four partial derivatives through mechanical tests $A = \frac{\partial f_1}{\partial p}$, $B = \frac{\partial f_1}{\partial q}$, $C = \frac{\partial f_2}{\partial p}$, $D = \frac{\partial f_2}{\partial q}$. For different stress paths, this book uses Gaussian fitting method to obtain the stress–strain relationship of the whole stress field (p, q) from the triaxial test data set, so as to obtain the values of four partial derivatives of each point in the stress field. They are expressed in the form of function series:

$$f(p, q) = \sum_{k=1}^{n} \omega(k)exp\{-a[(p - p_k)^2 + (q - q_k)^2]\} \tag{3.8}$$

where, $\exp\{-\alpha[(p - p_k)^2 + (q - q_k)^2]\}$ is a Gaussian function;$\omega(k)$ is the weight on the kth Gaussian center point;α is the adjustment parameter;p_k, q_k are the coordinates of Gaussian center point in (p, q) stress field;n is the total number of Gaussian centers.

3.3.2 Advantages of Numerical Modeling Method

Numerical modeling method has the following advantages.

(1) The numerical modeling method fully reflects the interaction between plastic volume strain and plastic shear strain of rock and soil, so it can comprehensively and accurately describe the basic characteristics of rock and soil deformation: compression, dilatancy and stress path correlation.
(2) Because the stress–strain relationship can be directly extracted from the triaxial test data and almost all the data obtained from the test can be fully utilized, the accuracy of the model obtained is high, and it has strong fault tolerance function.
(3) The stress path can be selected freely in modeling, which can simulate the actual stress path in geotechnical engineering.
(4) The determination of coefficient function in the model and the whole modeling process are completed by computer.

There are three preconditions for the generation of numerical modeling methods.

(1) With the development of geotechnical test technology, multi stress path triaxial apparatus and true triaxial apparatus have appeared and applied, which provide good test conditions for numerical modeling.
(2) With the increasing of computer capacity and computing speed, it provides a fast and convenient processing method for numerical modeling.
(3) The development of numerical analysis methods provides a wealth of high-precision numerical fitting and analysis methods.

With the rapid development of test technology, computer and numerical analysis method, numerical modeling method has a broad development prospect and can provide accurate constitutive model for computer simulation of geotechnical engineering.

3.4 Theory and Principle of Neural Network

The establishment of geotechnical constitutive model is actually an inverse problem, which is to deduce the constitutive relationship from the test data of soil, that is, to infer the cause from the results. For a long time, the inverse problem theory of geotechnical constitutive relation has been developing slowly. The main reason is the lack of effective inversion tools. It was not until the emergence, development and maturity of neural network, it is possible to solve this problem.

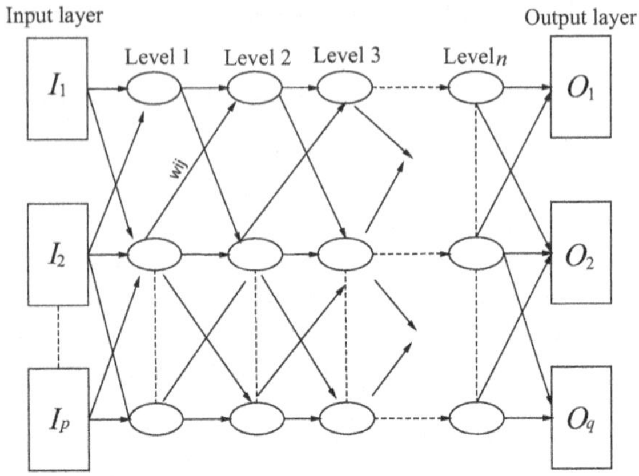

Fig. 3.2 Generalized neural network structure diagram

3.4.1 Structure of Neural Network

Artificial neural network (Ann) is a new information processing system which is developed on the basis of modern neuroscience research achievements and according to the basic functional characteristics of human brain, trying to simulate the function or structure of biological neural system. It is not a copy of the real biological nervous system, but only its mathematical abstraction and rough approximation and imitation. In essence, it is a kind of dynamic information processing system composed of a large number of basic information processing units through extensive connection. The structure of neural network mainly refers to its connection mode. Figure 3.2 shows the general structure of neural network.

3.4.2 Characteristics of Neural Network

Artificial neural network has absorbed many advantages of biological neural network and has the following characteristics.

(1) High degree of parallelism. Artificial neural network is composed of many same simple processing units in parallel. Although the function of each unit is simple, the parallel activities of a large number of simple processing units make its information processing ability and effect amazing.
(2) Highly nonlinear global action. Each neuron of artificial neural network receives a large number of inputs from other neurons, and influences other neurons through parallel network output. This mutual restriction and interaction between the network realizes the nonlinear mapping from the input state to the output state space. From the global point of view, the overall network performance is not a simple superposition of local network performance, but a collective behavior.

(3) Good fault tolerance and associative memory ability. Artificial neural network can realize the memory of information through its own network structure, and the memorized information is stored in the weights between neurons. The information content cannot be seen from a single weight, but is stored in a distributed way. The absence of information from one or a few neurons will not affect the entire network, which makes the network fault-tolerant and capable of processing pattern information such as cluster analysis, feature extraction and defect pattern recovery. It is also easy to perform pattern recognition such as pattern classification and pattern association.
(4) Strong adaptive and self-learning ability. Artificial neural network can obtain the weight and structure of the network through training and learning, which shows strong self-adaptive and self-learning ability.

Neural network is mainly used to reflect the computational model of learning and information processing, and self-learning is an important functional characteristic of neural network. It learns through the interaction of examples or the surrounding environment, and can learn complex nonlinear relations from large amounts of data.

3.4.3 *Learning Algorithm of Neural Network*

The learning of neural network is divided into directed and unsupervised. Guided learning requires that input and output target pairs be provided to the network during the learning period. At present, the application of neural network in geotechnical engineering is guided learning. This is because in the engineering research, we are based on the test data first. So the emphasis here is on learning by sample.

The basic idea of learning with samples is shown in Fig. 3.3. The purpose of learning is to generate a network (or more generally, a system). As shown in Fig. 3.3b, the network implements an unknown mapping $f: x \to y$. Given sufficient input and output pairs (x_i, y_i), the learning rules of a network define how to modify the weight when the expected output is generated from the given input. The learning process

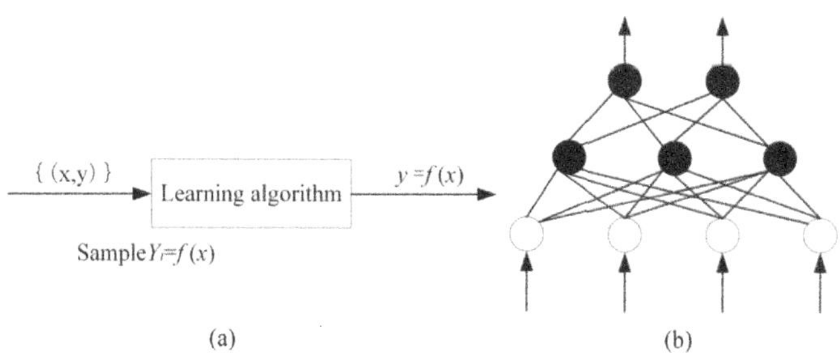

Fig. 3.3 Sample learning process

will modify the connection weight of the network by the learning algorithm, so as to obtain a good approximation of f.

There are many neural network algorithms, and the more mature algorithms are back propagation (BP) algorithm, radial basis function (RBF) algorithm, genetic algorithm (GA) and simulated annealing algorithm, etc.

3.5 Comparison of BP and RBF Neural Networks

3.5.1 BP Neural Network

BP neural network is the most widely used neural network model. As shown in Fig. 3.4 (only two hidden layers are shown), it is constructed in a hierarchical structure, including an input layer, an output layer and one or more hidden layers.

The learning process of network includes forward propagation and back propagation. In the forward propagation process, the input information is processed layer by layer through the hidden layer, and then transmitted to the output layer, the state of each neuron only affects the state of the next layer of neurons. If the output layer can not get the desired output, it turns to the back-propagation process and returns the error signal along the original connection path, and minimize the error signal by modifying the weights of neurons in each layer. BP algorithm adopts the most common gradient descent algorithm among optimization methods, which aims to minimize the mean square error between the actual output and the expected sample output. Generally, S-type logical nonlinear functions are used:

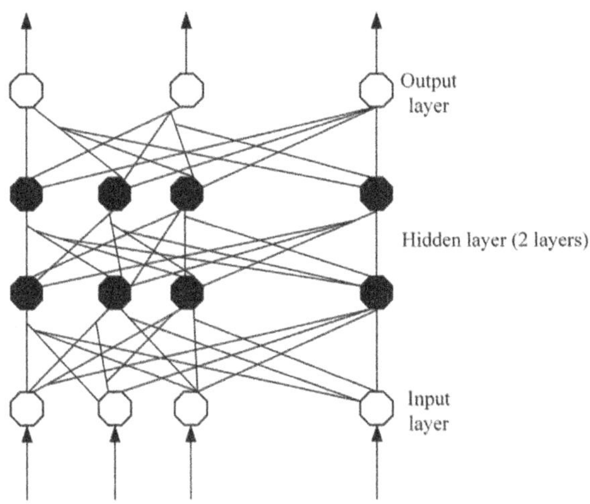

Fig. 3.4 BP neural network structure diagram

$$F(x) = \frac{1}{1 + \mathrm{e}^{-x}} \tag{3.9}$$

The specific algorithm is as follows (taking a hidden layer as an example, it can be carried out similarly for multiple hidden layers. Only the main formulas are listed here).

(1) Initializes weights W and thresholds θ by setting all weights and thresholds to smaller random numbers.
(2) Provide learning sample pairs (input and expected output values).
(3) Using S-type function and the following equations to calculate the output of all the hidden layer x'_j and output value of the output layer y_k. It is assumed that the hidden layer has n_1 units, the output layer has m units, and the input layer has n units:

$$x'_j = f\left(\sum_{i=0}^{n-1} w_{ij}x_i - \theta_j\right), 0 < j < n_1 - 1 \tag{3.10}$$

$$y_k = f\left(\sum_{j=0}^{n_1-1} w_{jk}x'_k - \theta_k\right), 0 < k < m - 1 \tag{3.11}$$

(4) Adjust weights and use recursive algorithm to reverse propagate errors from the output layer:

$$w_{ij}(t + 1) = w_{ij}(t) + \eta\delta_j x_j \tag{3.12}$$

where, η is the learning rate of the network and δ_j is the error term of node j.

BP neural network with three layers of Sigmoid nonlinear neurons can approach any continuous function with arbitrary precision, but there are some inherent shortcomings in BP neural network.

(1) From the point of view of mathematics, it is a nonlinear optimization problem, so there is inevitably the problem of local minimum.
(2) The convergence speed of its learning algorithm is slow, and it usually needs thousands or more iterations.
(3) There is no theoretical guidance for the selection of the number of hidden nodes in the network, but it is determined by trial and error based on experience.
(4) There is no certain criterion for the selection of learning parameters (step size and momentum term coefficient).

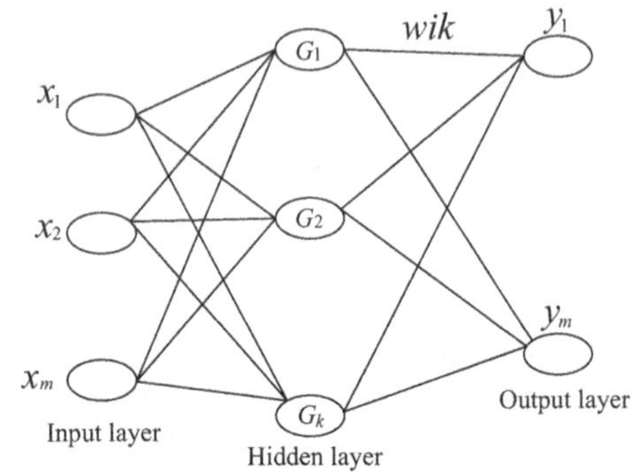

Fig. 3.5 RBF neural network structure diagram

3.5.2 RBF Neural Network

As shown in Fig. 3.5, RBF neural network consists of three layers: the first layer is the input layer, and the transfer function is RBF nonlinear function, which is a non negative nonlinear function with local distribution attenuating the central radial symmetry; the second layer is hidden layer, and the transfer function is linear function. The weight sum and threshold of each neuron in the hidden layer define the position and width of RBF function. A certain number of hidden layer neurons and the correct weight of each layer can approach any complex function and meet the design error requirements; the third layer is the output layer. On the whole, the input to output of the network is nonlinear, but from the hidden layer to the output layer is linear, so the weight of the network can be directly solved by the linear equations, which is also the reason why the RBF neural network has fast approximation speed and can avoid the local minimum problem.

The output of the kth node in the network output layer is a linear combination of the output of hidden nodes:

$$y_k = \sum_i w_{kj} q_i - \theta_k \tag{3.13}$$

where, θ_k is the threshold value of the kth output node.

3.5.3 RBF Learning Algorithm

The learning process of RBF neural network is to automatically generate one RBF neuron and adjust the corresponding weight value each time, so as to continuously increase the number of RBF neurons and adjust the weight until reaching the required error index and the maximum number of training steps. According to the above characteristics, for the input $x = \left(x_1, x_2, L, x_p\right)$, RBF function is usually selected is

Gaussian function, that is, the action function is

$$\alpha_j(x) = \exp\left(-\frac{\|x - c_j\|^2}{\sigma_j^2}\right) \quad (3.14)$$

where, c_j is the center point of the jth basis function, which can be determined by clustering analysis according to the input samples according to the self-organizing map neural network; and σ_j is the parameter that can be freely selected, which determines the width of the basis function around the center point; $\|x - c_j\|$ is the norm of $x - c_j$. So, with the increasing of $\|x - c_j\|$,$\alpha_j(x)$ quickly decay to zero, for a given input $x \in R^n$, only a small number of processing units whose center is close to X are activated, that is, to achieve the purpose of local approximation.

The learning algorithms of RBF neural network can be divided into two types: learning without tutor and learning with tutor. Because in geotechnical engineering, we only encounter the problem of having a tutor, so this book only introduces the learning algorithm with a tutor. Learning algorithms with mentors are also known as supervised learning. When c_j is determined, the weight system value between the hidden layer and the output layer can be determined, which is a system of linear equations, that is to say, to calculate the weight becomes a linear optimization problem, which can be obtained by various linear optimization algorithms, such as LMS algorithm, least square recurrence method, mirror mapping least square method, etc.

3.5.4 Comparison Between RBF Neural Network and BP Neural Network

One of the most important functions of neural network is to be able to effectively approximate complex nonlinear functions. There are two approaches: global approach and local approach. If one or more weights or adaptive adjustable parameters of the network affect any output at each point of the input space, it is called global approximation, and BP neural network is a typical example. For each input and output data, each weight of the network needs to be adjusted, so the global approximation is very slow. At the same time, due to the gradient descent method, there are some shortcomings such as local minimum and slow convergence speed. For a local region of the input space, if only a few weights affect the output of the network, it is called local approximation. As the local approximation neural network has only a small amount of weight adjustment for input and output data, it learns quickly, among which RBF neural network is a successful example. Table 3.1 shows the comparison between RBF neural network and BP neural network.

Table 3.1 Comparison of RBF neural network and BP neural network

Project	RBF neural network	BP neural network
Approximation way	Local approximation	A global approach
The rate of convergence	Fast, more than 1000 times faster than BP neural network, no local minimum	Slow, local minimum
The network structure	Three floors, fixed	Multiple layers, not fixed
Number of neurons in the hidden layer	Automatically adjusted by the program according to the calculation accuracy	artificial
Requirements for computer hardware	High, big memory	Low, small memory

3.6 Application of Neural Network in Geotechnical Engineering

In recent years, the development of neural network in geotechnical engineering mechanics is reflected in the coupling with other numerical methods. In most studies, the neural network is an independent system, but sometimes the neural network can be incorporated into the existing numerical methods (which is very difficult) for calculation. The key of the problem is how to integrate the neural network into the existing calculation methods, such as finite element method and FLAC and so on, so as to make the numerical analysis method more applicable.

Simulation is a basic method in the research of engineering problems. It can deepen the understanding of the phenomena and make it possible to carry out engineering analysis for the phenomena such as material constitutive characteristics. The neural network (Fig. 3.6) provides this new basic constitutive simulation tool.

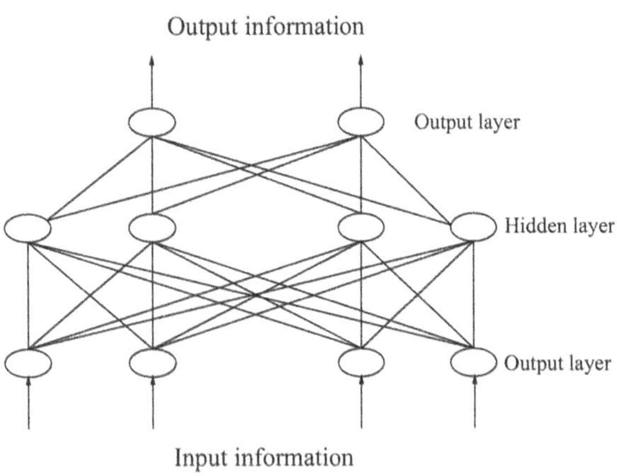

Fig. 3.6 Three-layer neural network constitutive model of soil

Here, the material properties (such as experimental data) should be measured in order to identify the main characteristics. Once the material characteristics are fully obtained, the main characteristics of the material can be identified, and the corresponding mathematical model can be established to simulate the characteristics. In Fig. 3.6, the artificial neurons in the input layer can represent stress state, strain state or stress increment. The neurons in the output layer can represent the strain increment. In this new constitutive simulation method, the constitutive characteristics can be directly obtained from structural tests by the learning ability of neural network and coded. In the training process, the activation values of input layer and output layer are specified, and the connection strength (or connection weight) is modified according to the learning rules. When the neural network is trained with a sufficient amount of data along different paths, it can give the correct strain increment for any set of stress increments under any stress state.

This new constitutive model provides a mapping from stress increment to strain increment (control stress model). However, in computational mechanics and finite element analysis, the inverse mapping from strain increment to stress increment is also necessary (control strain model). The important characteristic of neural network computing model is that inverse mapping is easy to obtain.

The training data used in the new neural network computing model must contain a sufficient number of stress paths as well as the knowledge of nonlinear mapping that can be acquired by the network. A set of data containing this knowledge is called a composite array. A neural network that has been trained with a synthetic array of constitutive properties can use its own inductive ability to give appropriate answers when looking for stress paths not included in the training array.

The theoretical research of neural network is a new frontier and interdisciplinary subject. Its emergence and development are influenced by other disciplines, which in turn will inevitably affect the development of other disciplines. The research content of artificial neural network is very rich, which includes five aspects: basic theory, model, algorithm, application and implementation. It is often used in prediction, analysis, optimization, control, diagnosis and identification, classification and identification. Therefore, it is of great significance to study neural networks.

Chapter 4
Triaxial Test and Numerical Modeling of Expansive Soil

4.1 Engineering Background and Physical Property Test

This book takes Lin huaigang flood control project as the background. The earth dam is located at the junction of Huo qiu County and Ying shang County in Anhui Province. It is a large flood control project in the middle reaches of Huai he River. The project is of first class and first class. The main dam is 7.7 m long, the maximum dam height is 17 m, the South auxiliary dam is 10.3 m long and the maximum dam height is 10 m. The north auxiliary dam is 61.75 m long and the maximum dam height is 8.5 m, all of which are designed as homogeneous dams. The construction of the main dam began in September 1958 and was stopped in 1962. During this period, the 4.7 m earth dam south of the main canal of Huai he River was basically completed. Due to the lack of any management and maintenance measures after the shutdown, the dam body is seriously damaged and many transverse cracks are found, with the maximum width of 5–6 cm.

1. **Mineral composition**

The mineral composition of the expansive soil in this test includes (clastic) minerals and clay minerals. By means of X-ray quantitative phase analysis, it was determined that the majority of granular minerals were quartz, accounting for 25–30%, followed by feldspar (potassium feldspar and sodium feldspar) accounting for 3–10%. The total amount of granular minerals in soil samples accounts for about 40%, and individual samples account for more than 50%. Using the same method, the total amount of clay minerals was determined to be greater than 50% and as high as 60%, among which the content of Montmorillonite and Mixed layer clay was about 30% and as high as 36%, and the content of illite was 5–12%, and the crystallization degree was not very good, as shown in Table 4.1 for details.

Q. Ren and J. Zhou, *Numerical Modeling of Soil Constitutive Relationship*,
https://doi.org/10.1007/978-981-16-3231-0_4

Table 4.1 Mineral composition of expansive soil

Number	Sampling depth (m)	Quartz (%)	K-feldspar (%)	Albite (%)	Montmorillonite (%)	Illite (%)	Kaolinite (%)
B1-1	0.6–0.9	30	5	8	30	16	11
B1-1*	0.6–0.9	30	5	9	28	16	12
B1-4	1.5–1.8	32	8	4	32	17	7
KB1	1.2–1.4	30	3	8	32	20	7
KB1*	1.2–1.4	44	8	2	25	15	6
KT1	1.2–1.5	30	3	8	30	18	11
B1-6	0.7–1.0	29	4	8	32	21	6
Bmix	0.9–1.2	30	3	9	33	20	5

Note The mark with * means compacted soil

2. **Chemical composition**

The chemical composition of expansive soil is mainly SiO_2, Al_2O_3, Fe_2O_3 three kinds of oxides, the total proportion of which exceeds 85%, see Table 4.2. It can be seen from the table that the silicon-aluminum molecular ratio of expansive soil is greater than 4.0, indicating that the clay mineral composition is mainly montmorillonite, which is consistent with the mineral identification results.

3. **Cation exchange capacity**

Particle exchange adsorption in expansive soil is an important chemical property of clay minerals. Measuring the cation exchange performance of expansive soil can qualitatively identify the main clay mineral types that make up the soil, thereby evaluating the hydrophilicity, expansive soil and strength of the soil. The specific situation is shown in Table 4.3. It can be seen from the table that the amount of ion exchange does not change much, generally 20.08–30.16 mol/kg, the specific surface

Table 4.2 Chemical composition content of expansive soil (Unit: %)

Number	SiO2	Al2O3	Fe2O3	CaO	MgO	Na2O	K2O	Ignition loss	SiO2/Al2O3
B1-1	64.14	16.07	5.33	0.83	1.40	1.01	2.30	7.41	6.77
B1-1*	64.88	15.54	5.39	0.83	1.36	1.08	2.26	7.56	7.08
B1-4	63.85	16.24	5.16	1.14	1.64	0.98	2.46	7.91	6.65
KB1	63.92	16.24	5.33	0.94	1.60	1.00	2.40	7.74	6.69
KB1*	73.29	11.56	3.60	0.80	1.01	1.06	1.83	5.60	7.72
B1-6	64.51	16.00	4.84	1.03	1.39	0.87	2.12	7.95	6.80
Bmix	64.60	15.88	5.20	0.97	1.53	1.08	2.41	7.57	6.89
KT1	65.27	16.89	4.83	0.77	1.04	0.88	2.05	7.46	6.54

Note The mark with * means compacted soil

Table 4.3 Expansive soil cation exchange capacity

Number	Ion exchange capacity (mol/kg)	Specific surface area (m^2/g)	pH	HCO_3^- (mol/kg)	Organic matter content (g/kg)	SO_4^{2-} (mol/kg)
B1-1	27.48	208	8.28	4.78	6.68	<0.5
B1-1*	27.24	224	8.20	4.48	6.97	<0.5
B1-4	29.52	312	8.45	2.91	10.09	<0.5
KB1	28.72	239	8.23	5.08	4.38	<0.5
KB1*	20.08	148	8.47	1.20	8.81	<0.5
B1-6	30.16	263	8.53	1.20	9.16	<0.5
Bmix	28.28	263	8.48	1.20	7.22	<0.5

Note The mark with * means compacted soil

area is generally 148–312 m^2/g, the pH is 8.20–8.53, the organic matter content is 4.38–11.69 g/kg, and the HCO^- value is 1.20–5.08 mol/kg.

4. **Soluble Salt and Cement**

It can be seen from the table that the content of soluble base cations is 1.25–7.44 mol/kg, with an average of 2.51 mol/kg. Table 4.4 lists the soluble salt and main free oxide content of the expansive soil in this area. The KT1 sample has a relatively high soluble salt content of 7.44 mol/kg. The reason may be the shallow sampling depth. The main cements of expansive soil are free SiO_2, Al_2O_3, Fe_2O_3, among which the free SiO_2 is 3.09–4.16 g/kg, the free Al_2O_3 is 6.30–11.10 g/kg, and the free Fe_2O_3 content is higher, 11.87–19.87 g/kg. The presence of these cements can enhance the structural connection between the expansive soil particles and form a certain structural strength.

Table 4.4 Expansive soil cementing material

Number	Cementing material (g/kg)			Soluble salt (mol/kg)				Total base (mol/kg)
	SiO_2	Al_2O_3	Fe_2O_3	Na^+	K^+	Mg^{2+}	Ca^{2+}	
B1-1	4.16	8.58	19.87	0.91	0.12	0.29	0.12	1.44
B1-1*	4.02	7.70	19.39	0.83	0.09	0.24	0.09	1.25
B1-4	3.88	9.75	12.63	1.50	0.17	0.43	0.25	2.35
KB1	3.71	9.98	16.17	1.28	0.11	0.48	0.09	1.96
KB1*	3.09	6.30	15.99	1.14	0.14	0.42	0.14	1.84
B1-6	3.97	11.10	13.24	2.01	0.08	0.46	0.30	2.85
Bmix	3.13	7.66	17.48	1.33	0.13	0.45	0.12	2.03
KT1	3.43	7.81	11.87	6.73	0.20	0.61	0.26	7.44

Note The mark with * means compacted soil

Table 4.5 Description of typical electron microscope photos of expansive soil samples

The sample number	Photo id	Magnification	Instructions
B1-1	001	3600	Flaky illite and curved flaky montmorillonite show skeleton-like microstructure with pore size of 1–5 and porosity of over 30%
B1-1*	002	3000	Honeycomb microstructure, mainly clay minerals, micromontmorillonite and illite, pore size 1–5, porosity greater than 50%, with microfractures
B1-4	003	5400	The curved lamellar montmorillonite is petal-like and the pore size is 5–10
KB1	004	780	The soil has a cellular microstructure
KB1*	005	2600	The mineral composition is curved and lamellar montmorillonite and lamellar illite
B1-6	006	3200	The soil is lamellar microstructure, the pores are not fully developed, and the mineral composition is illite
Bmix	007	3600	Granular minerals (quartz, feldspar microcrystals) are distributed in the pores of the soil with a porosity of more than 50%
KT1	008	3600	Curved lamellar montmorillonite microcrystal

Note The mark with * means compacted soil

5. **Icrostructure Characteristics**

After the soil sample was naturally dried, it was boiled and ground into thin slices, and then the mineral distribution characteristics were observed under an optical microscope. At the same time, the soil sample was scanned by electron microscope, and electron microscope photos of different magnifications were taken. The results are shown in Table 4.5 and Fig. 4.1. It can be seen from the Fig. 4.1 that the morphology, size and distribution characteristics of clay minerals such as montmorillonite and illite, as well as the type of microstructure and the size, shape and microporosity of pores.

4.2 Basic Physical Property Test of Expansive Soil

In order to obtain the actual stress–strain parameters, in addition to selecting the appropriate mathematical model, the method for measuring these parameters must also be planned according to the engineering and soil conditions, that is, the selection of experimental instruments, the determination of sample preparation methods, and the clear test Steps, etc., these test details have a greater impact on the test results.

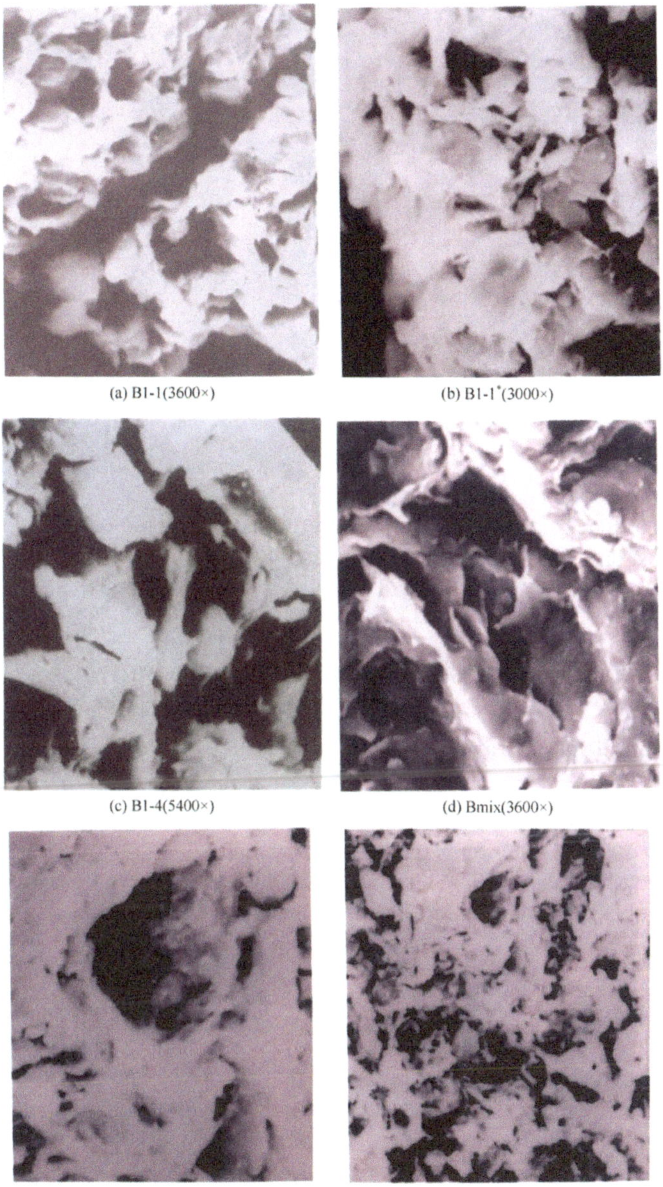

Fig. 4.1 Different magnification electron microscope photos of expansive soil

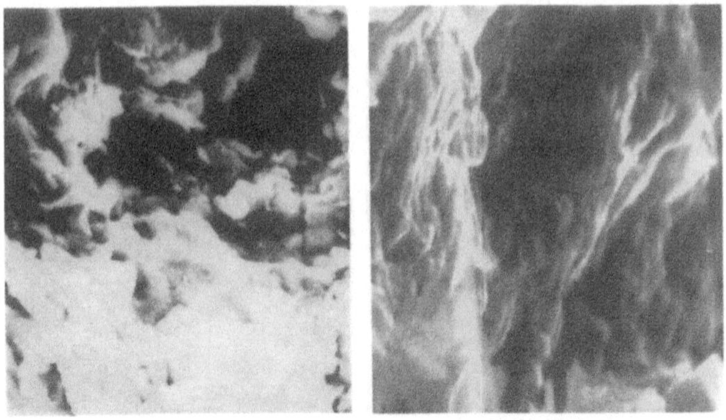

(g) KT1(3600×) (h) B1-6(3200×)

Fig. 4.1 (continued)

The physical index of the soil must be determined before the conventional triaxial test. The test data are shown in Tables 4.6 and 4.7.

4.3 Triaxial Compression Test of Expansive Soil

1. **The sample is saturated**

Since the saturation of the sample has a greater impact on the accuracy of the test results, the vacuum saturation method with better saturation effect is used. The pumping pressure reaches 1 atmosphere, and the sample is saturated for 48 h and then taken out. The saturation of the sample can reach about 95%.

2. **Sample cutting**

(1) Generally speaking, for the consolidated drained shear test, the shear rate should be controlled within 0.003–0.012% per minute, and for the consolidated undrained shear test, the shear rate should be controlled within 0.05–0.1%. After many tests, under the premise of not affecting the shear effect, the consolidation undrained shear test is selected.
(2) When the dynamometer reading shows a peak, shearing should continue until the axial strain exceeds 5%. When there is no peak in the dynamometer reading, the shear should be carried out until the axial strain is 15–20%.

3. **Triaxial test results of samples in different conditions**

Triaxial saturated consolidation drainage and undrained tests were carried out on the above eight soil samples. The details are as follows: The triaxial saturated consolidation drainage test with sample numbers B1-4, KB1 and KT1 was carried out; the

Table 4.6 Determination of wet density and moisture content of samples

Soil sample number	Instrument number	Expansion of the front ring knife and the quality of the soil (g)	After expansion of the ring knife and the quality of the soil (g)	Ring knife weight (g)	Dry soil weight (g)	Load (kPa)	Pressure (kPa)	W_{before} (%)	W_{after} (%)
B1-1	D1	172.27	174.88	54.76	96.98	1715	0.56	21.2	24.2
B1-1*	D2	172.36	175.18	54.7	97.04	2160	0.71	21.2	24.2
B1-4	D3	171.18	172.67	55.06	93.02	240	0.08	24.8	26.4
KB1	D4	172.42	174.11	54.67	94.65	180	0.06	24.4	26.2
KB1*	D5	168.04	172.33	49.7	93.93	1520	0.51	26.0	30.6
B1-6	D6	171.03	174.29	53.0	96.70	1622	0.54	22.1	25.4
Bmix	D7	166.06	170.96	50.7	92.14	1550	0.52	25.2	30.5
KT1	D8	173.63	176.04	53.1	99.60	1625	0.54	21.0	23.4

Note The mark with * means compacted soil

Table 4.7 Combined test of liquid and plastic limit

Sample No	Cone sinking depth h (mm)	The mass of wet soil (g)	The mass of dry soil (g)	Water content (%)	Liquid limit (%)	Plastic limit (%)	The plastic index	2 mm water content in depth of subsidence			
								The first point (%)	The second point (%)	The average (%)	Error (%)
B1-1	4.40	9.37	7.22	29.8	49.7	21.7	28.0	21.32	20.65	20.98	0.67
	10.25	16.38	11.64	40.7							
	15.60	13.21	8.92	48.1							
B1-1*	5.60	17.55	13.99	25.5	32.8	20.0	12.8	20.17	19.85	20.01	0.33
	12.90	16.69	12.77	30.7							
	19.55	16.61	12.41	33.8							
B1-4	4.75	10.64	8.17	30.2	47.6	21.4	26.2	20.59	20.26	20.42	0.33
	10.90	14.72	10.51	40.1							
	16.60	12.32	8.37	47.2							
KB1	3.30	15.17	11.95	26.9	49.2	22.2	27.0	22.42	21.97	22.19	0.44
	11.10	17.77	12.52	41.9							
	16.30	15.53	10.46	48.5							
KB1*	4.80	12.06	9.82	22.8	32.2	17.3	14.9	17.99	16.58	17.27	1.41
	9.40	17.19	13.55	26.8							
	16.10	16.88	12.82	31.7							
B1-6	5.10	15.82	12.18		22.0	14.9	22.3	22.00	22.04	22.02	0.05
	11.40	17.83	12.84	44.3							
	17.40	17.06	11.80								

(continued)

Table 4.7 (continued)

Sample No	Cone sinking depth h (mm)	The mass of wet soil (g)	The mass of dry soil (g)	Water content (%)	Liquid limit (%)	Plastic limit (%)	The plastic index	2 mm water content in depth of subsidence			
								The first point (%)	The second point (%)	The average (%)	Error (%)
Bmix	4.30	22.88	18.63	22.8	31.3	18.5	12.8	19.09	17.94	18.51	1.15
	10.70	21.75	17.03	27.7							
	17.20	25.02	19.05	31.4							
KT1	5.80	21.69	15.23	33.7	53.4	19.8	33.6	21.42	18.28	19.78	3.14
	9.50	21.45	15.33	40.0							
	15.54	21.12	13.99	51.0							

Note The mark with * means compacted soil

triaxial saturated consolidation undrained test with sample numbers B1-1, B1-6 and Bmix. Due to the limited space, only triaxial tests with confining pressure of 200 kPa under drainage condition are listed. The test data are shown in Tables 4.8, 4.9 and 4.10. Note that adding black data in the table indicates the start of unloading.

4.4 Triaxial Test Curve of Expansive Soil

Due to space limitations, only the stress–strain relationship curve of the sample under the drained condition at a confining pressure of 200 kPa and the stress–strain relationship curve under the undrained condition are listed, as shown in Figs. 4.2, 4.3, 4.4, 4.5, 4.6, 4.7, 4.8, 4.9, 4.10 and 4.11.

4.5 Neural Network Learning and Prediction

This section uses RBF neural network. The learning rule adopted by the RBF neural network is the Delta rule. The main point is to change the connection weight between units to reduce the error between the actual output of the system and the expected output. This rule is also called the Widrow-Hoff learning rule, that is, the minimum mean square error rule. Nielsen has proved that when each node has different thresholds, a hidden layer network can be used to approximate a continuous function in any closed interval. Therefore, a three-layer neural network based on the RBF algorithm can complete arbitrary n-dimensional to m-dimensional mapping. Therefore, a three-layer RBF neural network is selected to realize the function of stress–strain.

4.5.1 Input Layer and Output Layer Design

The input layer acts as a buffer memory, adding data sources to the network. The number of nodes depends on the dimensionality of the data source, that is, this node can represent each data source. Therefore, the most difficult design criterion is to figure out the correct data source. If there is a large amount of unprocessed or false information in the data source, it will definitely hinder the correct training of the network. Therefore, to eliminate those useless data and determine the appropriate number of data sources, roughly the following four steps are required.

(1) Determine the application-related data.
(2) Eliminate data sources that are technically and economically incompatible with actual data.
(3) Eliminate edge or unreliable data sources.

Table 4.8 Test B1-4 triaxial test data (confining pressure 200 kPa)

Axial deformation (μm)	Force measuring ring (mm)	Drain reading (mm)	Pore pressure (kPa)	Axial strain (%)	Corrected area (mm^2)	Principal stress difference (kPa)	Displacement (mL)	Major principal stress (kPa)	P (kPa)	Q (kPa)	Volume strain (%)	Shear strain (%)
30	6.1	27.1	5.00	0.24	30.32	0.53	0.00	2.53	2.18	0.53	0.00	0.24
55	13.5	27	5.00	0.44	30.37	1.16	0.10	3.16	2.39	1.16	0.03	0.43
76	17.2	27	5.00	0.61	30.42	0.30	0.10	2.30	2.10	1.48	0.03	0.60
121	35.9	26.9	7.00	0.96	30.52	3.07	0.20	5.07	3.02	3.07	0.05	0.95
131	73.2	26.7	7.00	1.04	30.53	6.26	0.40	8.26	4.09	6.26	0.10	1.01
147	108	26.6	9.00	1.17	30.56	9.22	0.50	11.22	5.07	9.22	0.13	1.13
157	125	26.5	9.00	1.25	30.58	10.67	0.60	12.67	5.56	10.67	0.16	1.20
165	135	26.35	16.00	1.31	30.60	11.51	0.60	13.51	5.84	11.51	0.16	1.26
175	146.8	26.2	16.00	1.39	30.60	12.52	0.90	14.52	6.17	12.52	0.23	1.32
186	157.3	26.1	18.00	1.48	30.62	13.41	1.00	15.41	6.47	13.41	0.26	1.39
203	172.2	25.9	19.00	1.62	30.65	14.67	1.20	16.67	6.89	14.67	0.31	1.51
215	180.4	25.8	20.00	1.71	30.67	15.35	1.30	17.35	7.12	15.35	0.34	1.60
223	185.5	25.8	20.00	1.78	30.69	15.78	1.30	17.78	7.26	15.78	0.34	1.66
233	193	25.7	21.00	1.86	30.70	16.41	1.40	18.41	7.47	16.41	0.36	1.73
350	200	25.5	22.00	2.79	30.98	16.85	1.60	18.85	7.62	16.85	0.41	2.65
373	210.8	25.3	24.00	2.97	31.02	17.73	1.80	19.73	7.91	17.73	0.46	2.82
392	218.3	25.15	22.00	3.12	31.06	18.34	1.95	20.34	8.11	18.34	0.50	2.95
410	224.9	25	21.00	3.26	31.09	18.88	2.10	20.88	8.29	18.88	0.54	3.09
423	229.2	24.9	22.00	3.37	31.12	19.22	2.20	21.22	8.41	19.22	0.56	3.18

(continued)

Table 4.8 (continued)

Axial deformation (μm)	Force measuring ring (mm)	Drain reading (mm)	Pore pressure (kPa)	Axial strain (%)	Corrected area (mm^2)	Principal stress difference (kPa)	Displacement (mL)	Major principal stress (kPa)	*P* (kPa)	*Q* (kPa)	Volume strain (%)	Shear strain (%)
430	231.5	24.85	22.00	3.42	31.13	19.41	2.25	21.41	8.47	19.41	0.58	3.23
470	243	24.6	22.00	3.74	31.22	20.32	2.50	22.32	8.77	20.32	0.64	3.53
481	245.5	24.6	22.00	3.83	31.24	20.51	2.50	22.51	8.84	20.51	0.64	3.62
495	248.5	24.5	22.00	3.94	31.27	20.74	2.60	22.74	8.91	20.74	0.66	3.72
510	252.9	24.4	21.00	4.06	31.30	21.09	2.70	23.09	9.03	21.09	0.69	3.83
533	259	24.2	20.00	4.24	31.35	21.57	2.90	23.57	9.19	21.57	0.74	4.00
575	268	24	20.00	4.58	31.44	22.25	3.10	24.25	9.42	22.25	0.79	4.32
597	272.5	23.9	19.00	4.75	31.49	22.59	3.20	24.59	9.53	22.59	0.81	4.48
611	275.6	23.8	19.00	4.86	31.52	22.82	3.30	24.82	9.61	22.82	0.83	4.59
643	282	23.65	19.00	5.12	31.59	23.30	3.45	25.30	9.77	23.30	0.87	4.83
660	286.5	23.5	19.00	5.25	31.63	23.64	3.50	25.64	9.88	23.64	0.88	4.96
690	290.8	23.4	19.00	5.49	31.71	23.94	3.55	25.94	9.98	23.94	0.89	5.20
712	294.3	23.35	19.00	5.67	31.76	24.18	3.60	26.18	10.06	24.18	0.90	5.37
728	296.8	23.3	19.00	5.80	31.80	24.36	3.65	26.36	10.12	24.36	0.91	5.49
733	298.3	23.25	19.00	5.84	31.80	24.48	3.80	26.48	10.16	24.48	0.95	5.52
750	300.2	23.2	19.00	5.97	31.83	24.62	4.00	26.62	10.21	24.62	1.00	5.64
762	301.8	23.15	18.00	6.07	31.85	24.73	4.10	26.73	10.24	24.73	1.02	5.73
782	305.2	23.05	19.00	6.23	31.90	24.97	4.20	26.97	10.32	24.97	1.05	5.88
802	309.9	23	16.00	6.39	31.94	25.32	4.30	27.32	10.44	25.32	1.07	6.03

(continued)

Table 4.8 (continued)

Axial deformation (μm)	Force measuring ring (mm)	Drain reading (mm)	Pore pressure (kPa)	Axial strain (%)	Corrected area (mm^2)	Principal stress difference (kPa)	Displacement (mL)	Major principal stress (kPa)	P (kPa)	Q (kPa)	Volume strain (%)	Shear strain (%)
825	313.5	22.9	16.00	6.57	32.00	25.57	4.35	27.57	10.52	25.57	1.08	6.21
845	316.2	22.8	16.00	6.73	32.05	25.75	4.40	27.75	10.58	25.75	1.09	6.36
865	319.6	22.7	19.00	6.89	32.10	25.99	4.50	27.99	10.66	25.99	1.12	6.51
886	322.8	22.6	18.00	7.05	32.15	26.20	4.55	28.20	10.73	26.20	1.13	6.68
906	326.3	22.6	18.00	7.21	32.20	26.45	4.60	28.45	10.82	26.45	1.14	6.83
941	332.8	22.45	15.00	7.49	32.30	26.90	4.65	28.90	10.97	26.90	1.15	7.11
964	336	22.4	15.00	7.68	32.36	27.10	4.70	29.10	11.03	27.10	1.16	7.29
987	339.8	22.35	15.00	7.86	32.42	27.36	4.75	29.36	11.12	27.36	1.17	7.47
1001	341.7	22.3	15.00	7.97	32.45	27.48	4.80	29.48	11.16	27.48	1.18	7.58
1028	344.5	22.2	14.00	8.18	32.52	27.65	4.90	29.65	11.22	27.65	1.20	7.78
1043	346.6	22.15	13.00	8.30	32.56	27.79	4.95	29.79	11.26	27.79	1.21	7.90
1067	350	22.1	13.00	8.50	32.62	28.00	5.00	30.00	11.33	28.00	1.22	8.09
1100	354	22	13.00	8.76	32.71	28.25	5.10	30.25	11.42	28.25	1.24	8.34
1133	356.6	21.9	13.00	9.02	32.79	28.38	5.20	30.38	11.46	28.38	1.26	8.60
1150	367.2	21.9	13.00	9.16	32.84	29.18	5.20	31.18	11.73	29.18	1.26	8.74
1175	359.8	21.85	12.00	9.36	32.91	28.54	5.25	30.54	11.51	28.54	1.27	8.93
1190	360.6	21.8	12.00	9.47	32.95	28.57	5.30	30.57	11.52	28.57	1.28	9.05
1210	362.5	21.8	11.00	9.63	33.00	28.67	5.30	30.67	11.56	28.67	1.28	9.21
1233	364	21.7	11.00	9.82	33.06	28.73	5.40	30.73	11.58	28.73	1.30	9.38

(continued)

Table 4.8 (continued)

Axial deformation (μm)	Force measuring ring (mm)	Drain reading (mm)	Pore pressure (kPa)	Axial strain (%)	Corrected area (mm^2)	Principal stress difference (kPa)	Displacement (mL)	Major principal stress (kPa)	P (kPa)	Q (kPa)	Volume strain (%)	Shear strain (%)
1250	365.4	21.7	11.00	9.95	33.11	28.80	5.40	30.80	11.60	28.80	1.30	9.52
1270	366.8	21.7	11.00	10.11	33.17	28.86	5.40	30.86	11.62	28.86	1.30	9.68
1317	369	21.6	12.00	10.49	33.30	28.92	5.50	30.92	11.64	28.92	1.31	10.05
1350	372	21.5	10.00	10.75	33.39	29.08	5.60	31.08	11.69	29.08	1.34	10.30
1375	375	21.45	10.00	10.95	33.46	29.25	5.65	31.25	11.75	29.25	1.34	10.50
1395	377	21.4	10.00	11.11	33.52	29.36	5.70	31.36	11.79	29.36	1.35	10.66
1420	380.2	21.3	10.00	11.31	33.58	29.55	5.80	31.55	11.85	29.55	1.38	10.85
1443	384.2	21.3	9.00	11.49	33.65	29.80	5.80	31.80	11.93	29.80	1.37	11.03
1465	387.8	21.25	9.00	11.66	33.71	30.02	5.85	32.02	12.01	30.02	1.38	11.20
1497	390.8	21.2	8.00	11.92	33.81	30.17	5.90	32.17	12.06	30.17	1.39	11.46
1514	391.9	21.2	8.00	12.05	33.86	30.21	5.90	32.21	12.07	30.21	1.39	11.59

Note Sample number: B1-4; Test date: July 13, 2002; Calibration factor of force ring: 0.261 kg/0.01 m; Initial height: 12.5 cm; The initial reading of the drain pipe: 27.1 cm^3; Loading rate: 0.011 mm/min

Table 4.9 Test KB1 triaxial test data (confining pressure 200 kPa)

Axial deformation (μm)	Force measuring ring (mm)	Drain reading (mm)	Axial strain (%)	Displacement (mL)	Corrected area (mm^2)	Principal stress difference (kPa)	Major principal stress (kPa)	P (kPa)	Q (kPa)	Volume strain (%)	Shear strain (%)
0	0	14	0.00	0.00	30.19	0.00	2.00	2.00	0.00	0.00	0.00
25	2.7	14	0.20	0.00	30.25	1.64	3.64	2.55	1.64	0.00	0.20
53	5.4	14	0.42	0.00	30.32	3.29	5.29	3.10	3.29	0.00	0.42
83	12.2	3.95	0.66	10.05	29.58	7.43	9.43	4.48	7.43	2.66	0.22
89	21.1	13.9	0.71	0.10	30.40	12.85	14.85	6.28	12.85	0.03	0.70
97	30.1	13.85	0.78	0.15	30.42	18.33	20.33	8.11	18.33	0.04	0.76
107	38.5	13.8	0.86	0.20	30.44	23.45	25.45	9.82	23.45	0.05	0.84
115	43.5	13.7	0.92	0.30	30.45	26.49	28.49	10.83	26.49	0.08	0.89
142	55.1	13.5	1.14	0.50	30.50	33.56	35.56	13.19	33.56	0.13	1.09
162	60.3	13.3	1.30	0.70	30.53	36.72	38.72	14.24	36.72	0.19	1.23
172	62.4	13.25	1.38	0.75	30.55	38.00	40.00	14.67	38.00	0.20	1.31
200	67.2	13	1.60	1.00	30.60	40.92	42.92	15.64	40.92	0.26	1.51
220	70.1	12.9	1.76	1.10	30.64	42.69	44.69	16.23	42.69	0.29	1.66
285	78	12.5	2.28	1.50	30.77	47.50	49.50	17.83	47.50	0.40	2.15
335	80.9	12.1	2.68	1.90	30.87	49.27	51.27	18.42	49.27	0.50	2.51
357	82.2	12	2.86	2.00	30.91	50.06	52.06	18.69	50.06	0.53	2.68
401	85	11.7	3.21	2.30	31.00	51.77	53.77	19.26	51.77	0.61	3.00
435	87.1	11.5	3.48	2.50	31.07	53.04	55.04	19.68	53.04	0.66	3.26
475	89.4	11.3	3.80	2.70	31.16	54.44	56.44	20.15	54.44	0.72	3.56

(continued)

Table 4.9 (continued)

Axial deformation (μm)	Force measuring ring (mm)	Drain reading (mm)	Axial strain (%)	Displacement (mL)	Corrected area (mm^2)	Principal stress difference (kPa)	Major principal stress (kPa)	*P* (kPa)	*Q* (kPa)	Volume strain (%)	Shear strain (%)
508	91.2	11.2	4.06	2.80	31.24	55.54	57.54	20.51	55.54	0.74	3.82
536	91.9	11.1	4.29	2.90	31.30	55.97	57.97	20.66	55.97	0.77	4.03
577	94.6	10.9	4.62	3.10	31.39	57.61	59.61	21.20	57.61	0.82	4.34
583	95.4	10.85	4.66	3.15	31.40	58.10	60.10	21.37	58.10	0.83	4.39
595	95.6	10.8	4.76	3.20	31.43	58.22	60.22	21.41	58.22	0.85	4.48
615	97.2	10.75	4.92	3.25	31.48	59.19	61.19	21.73	59.19	0.86	4.63
660	99.8	10.55	5.28	3.45	31.58	60.78	62.78	22.26	60.78	0.91	4.98
674	100.5	10.5	5.39	3.50	31.62	61.20	63.20	22.40	61.20	0.93	5.08
690	101.6	10.5	5.52	3.50	31.66	61.87	63.87	22.62	61.87	0.93	5.21
710	102.5	10.4	5.68	3.60	31.70	62.42	64.42	22.81	62.42	0.95	5.36
740	104	10.3	5.92	3.70	31.78	63.34	65.34	23.11	63.34	0.98	5.59
800	107	10.1	6.40	3.90	31.92	65.16	67.16	23.72	65.16	1.03	6.06
841	110	9.8	6.73	4.20	32.01	66.99	68.99	24.33	66.99	1.11	6.36
884	111	9.75	7.07	4.25	32.12	67.60	69.60	24.53	67.60	1.13	6.70
907	112.1	9.75	7.26	4.25	32.19	68.27	70.27	24.76	68.27	1.13	6.88
931	113.1	9.7	7.45	4.30	32.25	68.88	70.88	24.96	68.88	1.14	7.07
972	115.1	9.6	7.78	4.40	32.36	70.10	72.10	25.37	70.10	1.17	7.39
1015	116.9	9.5	8.12	4.50	32.47	71.19	73.19	25.73	71.19	1.19	7.72
1085	120	9.3	8.68	4.70	32.65	73.08	75.08	26.36	73.08	1.25	8.26

(continued)

Table 4.9 (continued)

Axial deformation (μm)	Force measuring ring (mm)	Drain reading (mm)	Axial strain (%)	Displacement (mL)	Corrected area (mm^2)	Principal stress difference (kPa)	Major principal stress (kPa)	*P* (kPa)	*Q* (kPa)	Volume strain (%)	Shear strain (%)
1120	121.4	9.2	8.96	4.80	32.74	73.93	75.93	26.64	73.93	1.27	8.54
1163	123.1	9.1	9.30	4.90	32.86	74.97	76.97	26.99	74.97	1.30	8.87
1190	124.5	9	9.52	5.00	32.93	75.82	77.82	27.27	75.82	1.32	9.08
1277	128	8.95	10.22	5.05	33.18	77.95	79.95	27.98	77.95	1.34	9.77
1323	129.8	8.8	10.58	5.20	33.30	79.05	81.05	28.35	79.05	1.38	10.12
1366	131.5	8.7	10.93	5.30	33.42	80.08	82.08	28.69	80.08	1.40	10.46
1427	133.8	8.6	11.42	5.40	33.60	81.48	83.48	29.16	81.48	1.43	10.94
1461	135.2	8.5	11.69	5.50	33.69	82.34	84.34	29.45	82.34	1.46	11.20
1500	136.5	8.5	12.00	5.50	33.81	83.13	85.13	29.71	83.13	1.46	11.51
1551	138.2	8.4	12.41	5.60	33.96	84.16	86.16	30.05	84.16	1.48	11.91
1606	140.3	8.4	12.85	5.60	34.13	85.44	87.44	30.48	85.44	1.48	12.35
1635	141.3	8.35	13.08	5.65	34.22	86.05	88.05	30.68	86.05	1.50	12.58
1673	142.8	8.25	13.38	5.75	34.33	86.97	88.97	30.99	86.97	1.52	12.88
1712	144.1	8.2	13.70	5.80	34.45	87.76	89.76	31.25	87.76	1.54	13.18
1750	145.2	8.2	14.00	5.80	34.57	88.43	90.43	31.48	88.43	1.54	13.49
1785	146.3	8.15	14.28	5.85	34.68	89.10	91.10	31.70	89.10	1.55	13.76
1787	146.3	8.15	14.30	5.85	34.68	89.10	91.10	31.70	89.10	1.55	13.78
1789.5	143.9	8.2	14.32	5.80	34.69	87.64	89.64	31.21	87.64	1.54	13.80
1789.6	140	8.2	14.32	5.80	34.70	85.26	87.26	30.42	85.26	1.54	13.80

(continued)

Table 4.9 (continued)

Axial deformation (μm)	Force measuring ring (mm)	Drain reading (mm)	Axial strain (%)	Displacement (mL)	Corrected area (mm^2)	Principal stress difference (kPa)	Major principal stress (kPa)	*P* (kPa)	*Q* (kPa)	Volume strain (%)	Shear strain (%)
1789.7	135	8.2	14.32	5.80	34.70	82.22	84.22	29.41	82.22	1.54	13.81
1789	130	8.2	14.31	5.80	34.69	79.17	81.17	28.39	79.17	1.54	13.80
1788.5	125	8.2	14.31	5.80	34.69	76.13	78.13	27.38	76.13	1.54	13.80
1787	118.5	8.2	14.30	5.80	34.69	72.17	74.17	26.06	72.17	1.54	13.78
1785	110	8.15	14.28	5.85	34.68	66.99	68.99	24.33	66.99	1.55	13.76
1779	90	8.15	14.23	5.85	34.66	54.81	56.81	20.27	54.81	1.55	13.72
1774	80	8.2	14.19	5.80	34.64	48.72	50.72	18.24	48.72	1.54	13.68
1765	62.5	8.2	14.12	5.80	34.62	38.06	40.06	14.69	38.06	1.54	13.61
1752	46.3	8.2	14.02	5.80	34.57	28.20	30.20	11.40	28.20	1.54	13.50
1738	32.4	8.2	13.90	5.80	34.53	19.73	21.73	8.58	19.73	1.54	13.39
1728	20	8.25	13.82	5.75	34.50	12.18	14.18	6.06	12.18	1.52	13.32
1710	10.4	8.25	13.68	5.75	34.44	6.33	8.33	4.11	6.33	1.52	13.17
1696	4	8.25	13.57	5.75	34.40	2.44	4.44	2.81	2.44	1.52	13.06
1679	−1	8.3	13.43	5.70	34.35	−0.61	1.39	1.80	−0.61	1.51	12.93
1672	−1	8.3	13.38	5.70	34.33	−0.61	1.39	1.80	−0.61	1.51	12.87

Note Sample number: KB1-3; Test date: July 14, 2002; Calibration factor of force ring: 0.312 kg/0.01 m; Initial height: 12.47 cm; Initial reading of drain pipe: 14 cm^3; Loading rate: 0.011 mm/min

Table 4.10 Test KT1 triaxial test data (confining pressure 200 kPa)

Axial deformation (μm)	Force measuring ring (mm)	Pore pressure (kPa)	Drain reading (mm)	Axial strain (%)	Corrected area (mm^2)	Principal stress difference (kPa)	Displacement (mL)	Major principal stress (kPa)	*P* (kPa)	*Q* (kPa)	Volume strain (%)	Shear strain (%)
0	0	0	21.70	0.00	12.01	0.00	0.00	2.00	2.00	0.00	0.00	0.00
10	2.4	3	21.70	0.13	12.02	0.52	0.00	2.52	2.17	0.52	0.00	0.13
28	4.6	4	21.65	0.35	12.04	1.00	0.05	3.00	2.33	1.00	0.06	0.33
40	6.8	4	21.65	0.50	12.06	1.47	0.05	3.47	2.49	1.47	0.06	0.48
50	11	4	21.65	0.63	12.08	2.38	0.05	4.38	2.79	2.38	0.06	0.61
62	12	5	21.65	0.78	12.09	2.59	0.05	4.59	2.86	2.59	0.06	0.76
78	19.1	6	21.65	0.98	12.12	4.11	0.05	6.11	3.37	4.11	0.06	0.96
98	24.2	7	21.65	1.23	12.15	5.20	0.05	7.20	3.73	5.20	0.06	1.21
115	29	8	21.65	1.44	12.18	6.22	0.05	8.22	4.07	6.22	0.06	1.42
133	34.6	9	21.60	1.56	12.20	7.40	0.10	9.40	4.47	7.40	0.11	1.63
141	40.5	10	21.60	1.76	12.21	8.66	0.10	10.66	4.89	8.66	0.11	1.73
152	50.7	12	21.60	1.90	12.23	10.82	0.10	12.82	5.61	10.82	0.11	1.86
168	60.6	15	21.55	2.10	12.25	12.92	0.15	14.92	6.31	12.92	0.17	2.04
189	60.8	19	21.50	2.36	12.27	12.93	0.20	14.93	6.31	12.93	0.22	2.29
230	75.1	24	21.35	2.88	12.32	15.91	0.35	17.91	7.30	15.91	0.39	2.75
250	80.9	26	21.25	3.13	12.34	17.12	0.45	19.12	7.71	17.12	0.50	2.96
288	89.7	27	21.20	3.60	12.39	18.89	0.50	20.89	8.30	18.89	0.56	3.41
300	91.8	28	21.10	3.75	12.40	19.33	0.60	21.33	8.44	19.33	0.67	3.53
330	96.7	28	21.00	4.13	12.43	20.30	0.70	22.30	8.77	20.30	0.78	3.87

(continued)

Table 4.10 (continued)

Axial deformation (μm)	Force measuring ring (mm)	Pore pressure (kPa)	Drain reading (mm)	Axial strain (%)	Corrected area (mm^2)	Principal stress difference (kPa)	Displacement (mL)	Major principal stress (kPa)	P (kPa)	Q (kPa)	Volume strain (%)	Shear strain (%)
370	102.8	28	20.90	4.63	12.48	21.49	0.80	23.49	9.16	21.49	0.89	4.33
402	106.9	29	20.85	5.03	12.53	22.27	0.85	24.27	9.42	22.27	0.94	4.71
425	110.4	28	20.75	5.31	12.56	22.95	0.95	24.95	9.65	22.95	1.05	4.96
440	112.1	27	20.70	5.50	12.57	23.27	1.00	25.27	9.76	23.27	1.11	5.13
456	114.8	26	20.60	5.70	12.59	23.80	1.10	25.80	9.93	23.80	1.22	5.29
511	122.4	25	20.55	6.39	12.67	25.21	1.15	27.21	10.40	25.21	1.28	5.96
554	127.1	23	20.50	6.93	12.74	26.04	1.20	28.04	10.68	26.04	1.33	6.48
590	130.8	22	20.50	7.38	12.80	26.67	1.20	28.67	10.89	26.67	1.33	6.93
633	134.1	22	20.50	7.91	12.88	27.18	1.20	29.18	11.06	27.18	1.33	7.47
654	138.1	20	20.45	8.18	12.91	27.93	1.25	29.93	11.31	27.93	1.39	7.71
676	139.6	19	20.45	8.45	12.95	28.15	1.25	30.15	11.38	28.15	1.39	7.99
718	144.7	18	20.35	8.98	13.01	29.04	1.35	31.04	11.68	29.04	1.50	8.48
815	151.6	15	20.30	10.19	13.17	30.03	1.40	32.03	12.01	30.03	1.55	9.67
851	152.9	14	20.25	10.64	13.23	30.15	1.45	32.15	12.05	30.15	1.61	10.10
946	157.1	11	20.20	11.83	13.41	30.59	1.50	32.59	12.20	30.59	1.67	11.27
993	156.3	11	20.15	12.41	13.49	30.25	1.55	32.25	12.08	30.25	1.72	11.84
1115	156.3	10	20.15	13.94	13.73	29.72	1.55	31.72	11.91	29.72	1.72	13.36
1210	158.7	9	20.15	15.13	13.92	29.76	1.55	31.76	11.92	29.76	1.72	14.55
1230	158.6	8	20.15	15.38	13.96	29.65	1.55	31.65	11.88	29.65	1.72	14.80

(continued)

Table 4.10 (continued)

Axial deformation (μm)	Force measuring ring (mm)	Pore pressure (kPa)	Drain reading (mm)	Axial strain (%)	Corrected area (mm^2)	Principal stress difference (kPa)	Displacement (mL)	Major principal stress (kPa)	P (kPa)	Q (kPa)	Volume strain (%)	Shear strain (%)
1344	157.2	6	20.10	16.80	14.19	28.91	1.60	30.91	11.64	28.91	1.78	16.21
1613	156.8	4	20.10	20.16	14.79	27.67	1.60	29.67	11.22	27.67	1.78	19.57
1642	155.7	4	20.10	20.53	14.86	27.35	1.60	29.35	11.12	27.35	1.78	19.93
1670	155.7	4	20.10	20.88	14.92	27.23	1.60	29.23	11.08	27.23	1.78	20.28
1712	154.3	2	20.20	21.40	15.04	26.78	1.50	28.78	10.93	26.78	1.67	20.84
1712	**154.3**	**2**	**20.20**	**21.40**	**15.04**	**26.78**	**1.50**	**28.78**	**10.93**	**26.78**	**1.67**	**20.84**
1713	150	2	20.25	21.41	15.05	26.02	1.45	28.02	10.67	26.02	1.61	20.88
1713	143.2	2	20.25	21.41	15.05	24.84	1.45	26.84	10.28	24.84	1.61	20.88
1713	134	2	20.25	21.41	15.05	23.24	1.45	25.24	9.75	23.24	1.61	20.88
1713	115	2	20.25	21.41	15.05	19.95	1.45	21.95	8.65	19.95	1.61	20.88
1711.5	104	1	20.28	21.39	15.05	18.04	1.42	20.04	8.01	18.04	1.58	20.87
1708	87.5	1	20.28	21.35	15.04	15.18	1.42	17.18	7.06	15.18	1.58	20.82
1706	81.9	0	20.28	21.33	15.04	14.22	1.42	16.22	6.74	14.22	1.58	20.80
1703	72.3	0	20.28	21.29	15.03	12.56	1.42	14.56	6.19	12.56	1.58	20.76
1701	66	0	20.28	21.26	15.02	11.47	1.42	13.47	5.82	11.47	1.58	20.74
1699	61.5	0	20.28	21.24	15.02	10.69	1.42	12.69	5.56	10.69	1.58	20.71
1697	57	0	20.28	21.21	15.02	9.91	1.42	11.91	5.30	9.91	1.58	20.69
1694	51.1	0	20.28	21.18	15.01	8.89	1.42	10.89	4.96	8.89	1.58	20.65
1692	47.2	−1	20.28	21.15	15.00	8.21	1.42	10.21	4.74	8.21	1.58	20.62

(continued)

Table 4.10 (continued)

Axial deformation (μm)	Force measuring ring (mm)	Pore pressure (kPa)	Drain reading (mm)	Axial strain (%)	Corrected area (mm^2)	Principal stress difference (kPa)	Displacement (mL)	Major principal stress (kPa)	*P* (kPa)	*Q* (kPa)	Volume strain (%)	Shear strain (%)
1690	43.8	−1	20.28	21.13	15.00	7.62	1.42	9.62	4.54	7.62	1.58	20.60
1687	39.2	−2	20.28	21.09	14.99	6.82	1.42	8.82	4.27	6.82	1.58	20.56
1684	34.7	−2	20.28	21.05	14.98	6.04	1.42	8.04	4.01	6.04	1.58	20.52
1681	30.6	−3	20.28	21.01	14.98	5.33	1.42	7.33	3.78	5.33	1.58	20.49
1673	20.5	−3	20.28	20.91	14.96	3.58	1.42	5.58	3.19	3.58	1.58	20.39
1664	12.7	−4	20.30	20.80	14.94	2.22	1.40	4.22	2.74	2.22	1.55	20.28
1659	8.3	−5	20.35	20.74	14.94	1.45	1.35	3.45	2.48	1.45	1.50	20.24
1653	3.5	−5	20.35	20.66	14.92	0.61	1.35	2.61	2.20	0.61	1.50	20.16
1648.5	0	−5	20.35	20.61	14.91	0.00	1.35	2.00	2.00	0.00	1.50	20.11
1645	−1.9	−5	20.35	20.56	14.90	−0.33	1.35	1.67	1.89	−0.33	1.50	20.06
1641	−3	−5	20.35	20.51	14.89	−0.53	1.35	1.47	1.82	−0.53	1.50	20.01
1635	−3.9	−5	20.35	20.44	14.88	−0.68	1.35	1.32	1.77	−0.68	1.50	19.94

Note Sample number: KT1-3; Test date: July 17, 2002; Calibration factor of force ring: 0.261 kg/0.01 m; Initial height: 8.0 cm; Initial reading of the drain pipe: 21.7 cm^3; Loading rate: 0.011 mm/min

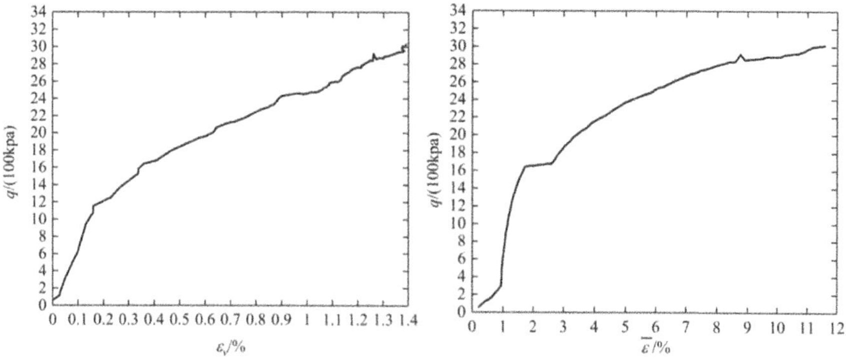

Fig. 4.2 Triaxial test data of sample B1-4 (confining pressure 200 kPa)

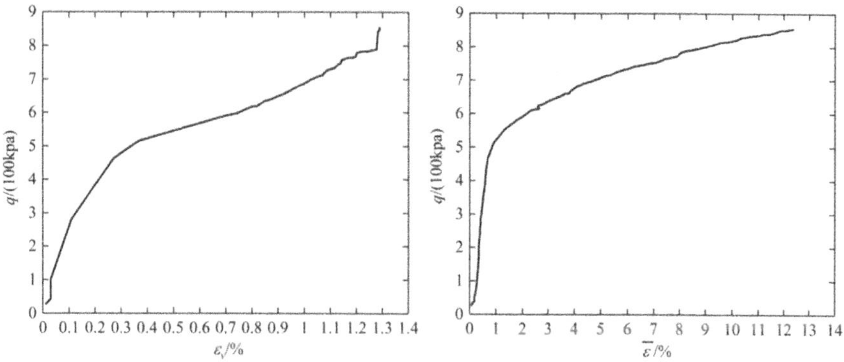

Fig. 4.3 Triaxial test data of sample KB1 (confining pressure 200 kPa)

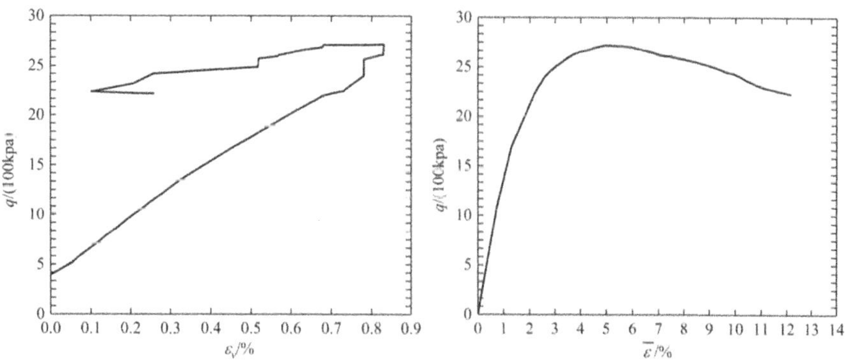

Fig. 4.4 Triaxial test data of sample KT1 (confining pressure 200 kPa)

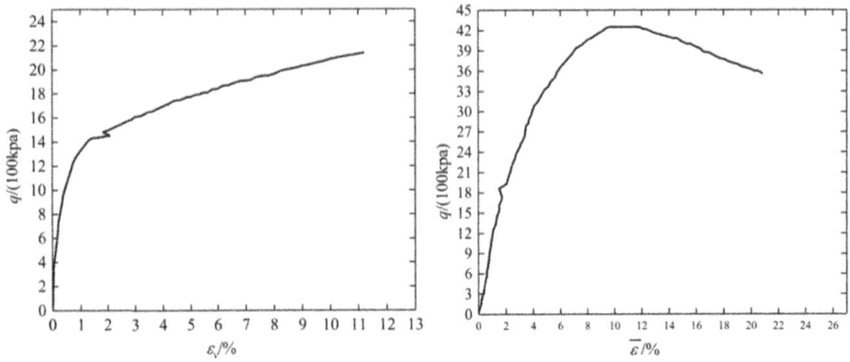

Fig. 4.5 Triaxial test data of sample B1-1 (confining pressure 100 kPa)

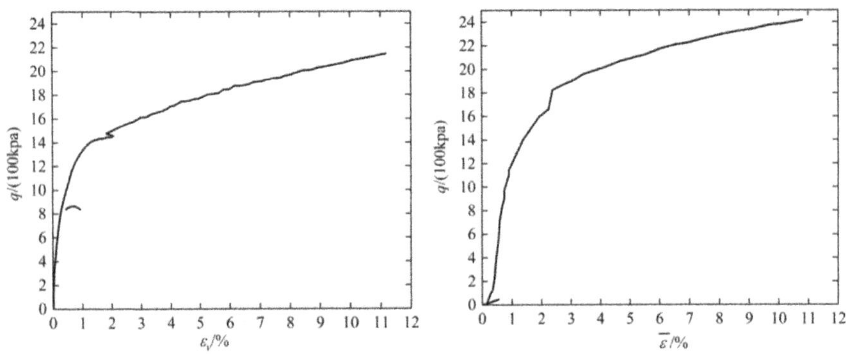

Fig. 4.6 Triaxial test data of sample B1-1 (confining pressure 200 kPa)

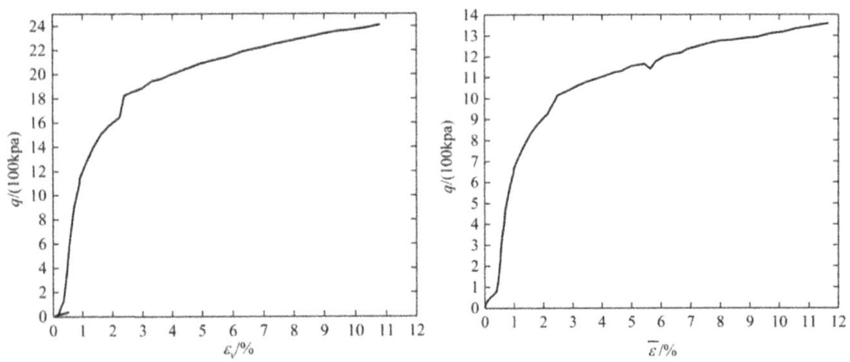

Fig. 4.7 Triaxial test data of sample B1-1 (confining pressure 300 kPa)

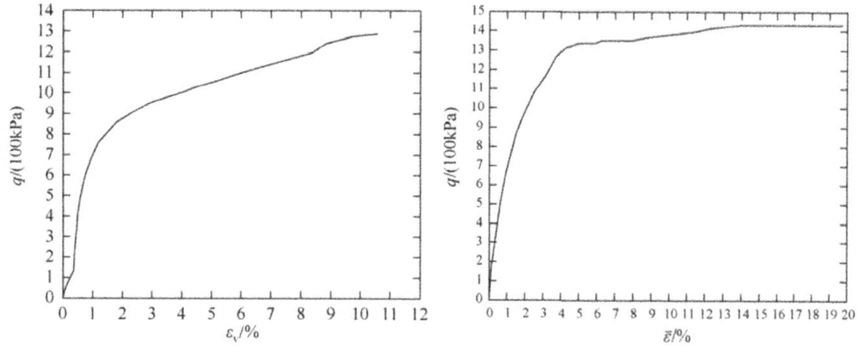

Fig. 4.8 Triaxial test data of sample B1-6 (confining pressure 100 kPa)

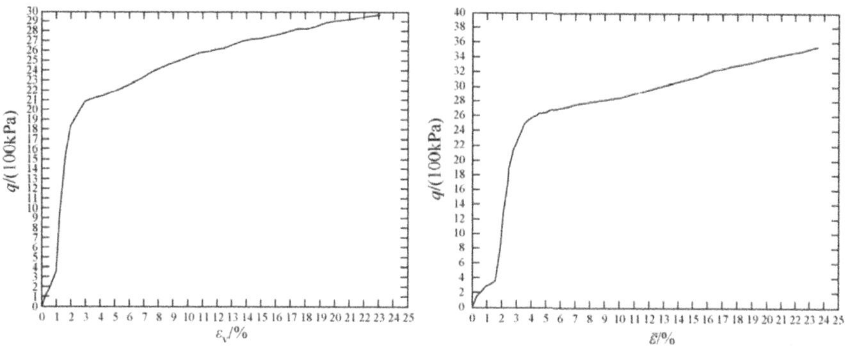

Fig. 4.9 Triaxial test data of sample B1-6 (confining pressure 200 kPa)

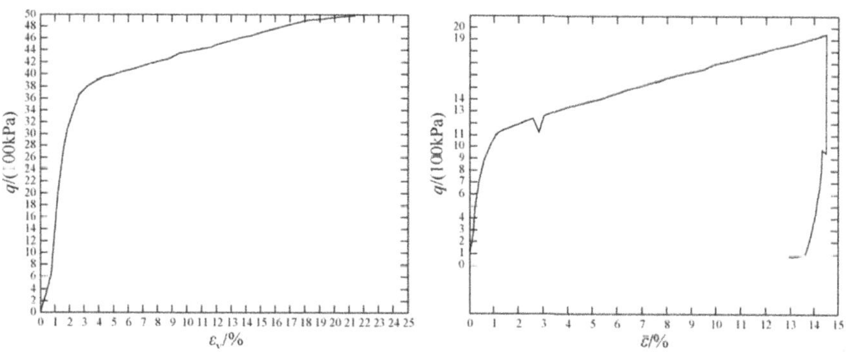

Fig. 4.10 Triaxial test data of sample Bmix (confining pressure 100 kPa)

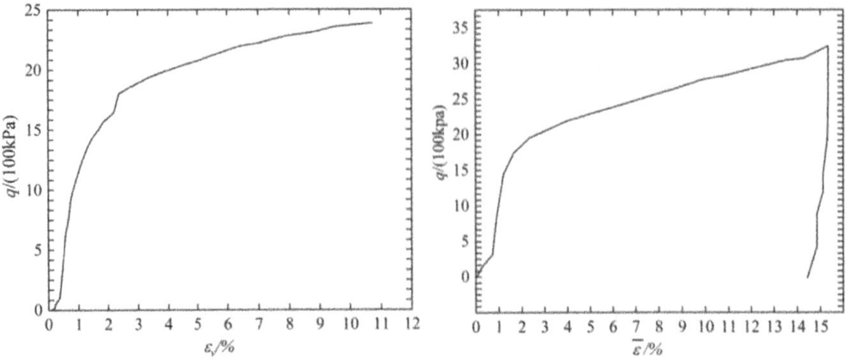

Fig. 4.11 Triaxial test data of sample Bmix (confining pressure 200 kPa)

(4) Develop a method can combine or preprocess data to make the data more practical.

Artificial neural networks can only process input and output data expressed as numerical values, so external information is often transformed into codes. For example, the external information is "male" and "female", represented by "1" or "0". When designing the input layer and output layer, the scale of the system should be reduced as much as possible to reduce the learning time and complexity of the system.

4.5.2 Selection of Hidden Layer Neurons

In this section, based on the triaxial test, the wrong and useless data are eliminated, and then the octahedral normal stress p and octahedral shear stress q are taken as the input parameters, and the volume strain ε_v and shear strain $\bar{\varepsilon}$ are taken as the output parameters. According to the designed stress path, the test data are selected as the samples of network training, and the test data are normalized before training.

The neural network generates a RBF neural network neuron every time the error is retransmitted, and the number of neurons is continuously increased until the error index or the maximum number of training steps is reached. This is also an advantage of RBF neural network compared with BP neural network: the hidden layer units in BP neural network are often determined by human according to experience, and it is difficult to accurately determine the number of hidden nodes, while the hidden nodes in RBF neural network are automatically determined by the network according to the training requirements.

4.5.3 RBF Algorithm

(1) Determine the center c_i of RBF. The center c_i of RBF, $c_i = (c_{i1}, c_{i2}, \ldots, c_{in})(i = 1, 2, \ldots, m)$. There are $m \times n$ parameters, which can be used to calculate c_i. The idea is to make the c_i sample the input data as evenly as possible, and the c_i is also dense where the data points are dense. The mean clustering method, which is widely used in pattern recognition, is used to determine its value. The specific calculation steps are as follows.

Step 1: Take m sample values from the learning samples as the initial value of c_i.
Step 2: Group all learning samples according to the nearest RBF center c_i. For every c_i and all the sample patterns $x^u = \left(x_1^u, x_2^u, \ldots, x_n^u\right)$ satisfies $\min\left(\sum_n^k \left(x_k^u - x_k^j\right)\right)$, x^u belongs to the subset θ_j of c_i.
Step 3: Recalculate $c_{jk} : c_{jk} = \frac{1}{m_j}\sum_{u\in\theta} x_k^u$, where m_j is the number of sample patterns in the subsample set θ_j.
Step 4: If c_i of RBF center is no longer changed, the training will be stopped, and the stable value of c_i will be calculated. Otherwise, turn to step 2 and recalculate c_{jk} until it is stable.

(2) Determine the width of RBF σ_j: after the training of RBF center c_i, we can obtain the normalized parameters, that is, the width of the basis function σ_j, which represents a measure of the sample dispersion in the sub sample set associated with each center, and can be determined by many methods. The most common method is to make them equal to the average distance between the RBF center and the sample pattern in the sub sample set, namely, $\sigma_j^2 = \frac{1}{m_j}\sum_{u\in\theta_j}\sum_{k=1}^n \left(x_k^u - x_{jk}\right)^2$.

(3) Action function: the most common action function is Gaussian function:

$$R_j(x) = \exp\left(-\frac{\|x - c_j\|^2}{\sigma_j^2}\right)$$

BP algorithm can still be used to modify the connection weight. Since $R_j(x)$ is a Gaussian function, it has $R_j(x) > 0$ for any x, thus losing the advantage of local adjustment weight. In fact, $R_j(x)$ is very small when x is far away from c_i, so it can be regarded as 0. Therefore, in fact, only when $R_j(x)$ is greater than a certain value can the corresponding weight be modified. The RBF neural network after such treatment also has the advantage of fast convergence of local approximation learning. At the same time, this approximation can overcome the shortcoming that Gaussian basis function does not have compactness to a certain extent.

4.5.4 Comparison Between RBF Neural Network and BP Neural Network

It can be seen from Figs. 4.12 and 4.13 that RBF neural network trains 71 groups of two-dimensional array data. After 47 steps of training, the network error meets the accuracy requirement (10^{-15}). While BP neural network trains the same batch of data, after 50,000 steps of training, the error is still about 10^0, which is far from the target. The calculation shows that when the error accuracy is set at 10^{-3}, the network training is trapped in the local minimum, which can not reach this accuracy.

There are also some problems in RBF neural network that need further study. This paper makes a comprehensive comparison between them.

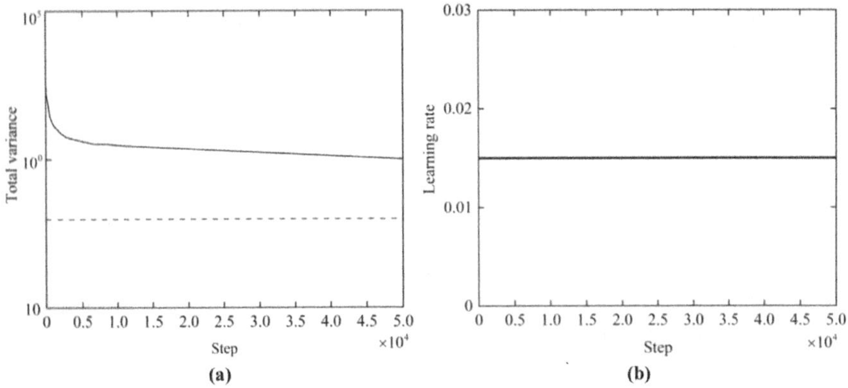

Fig. 4.12 Time step error of BP algorithm

Fig. 4.13 Time step error of RBF algorithm

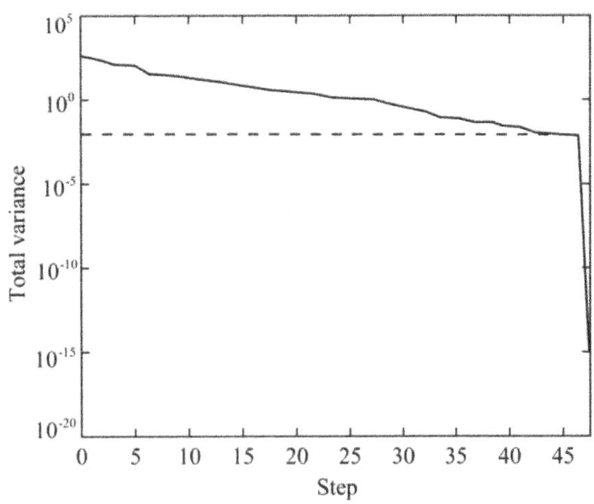

(1) Different action functions are used in nonlinear mapping. The action function of RBF is local and that of BP is global.
(2) It has been proved that RBF neural network has the characteristic of unique best approximation and no local minimum.
(3) It is a difficult problem to find the center c_i of RBF neural network node and to standardize the parameter σ_j^2.
(4) RBF neural network is used in the identification and control of nonlinear systems. Although it has the characteristics of unique best approximation and no local minimum, it is difficult to find the center of hidden node, which is the reason why RBF neural network is difficult to be widely used.

4.5.5 *Prediction Effect of RBF Neural Network*

The stress–strain relationship of six kinds of expansive soil samples under confining pressure of 200 kPa is predicted respectively. Here, the test data under confining pressure of 100 and 300 kPa are taken as learning samples. After training, they are used to predict the stress–strain situation of samples under confining pressure of 200 kPa. The results are shown in Figs. 4.14, 4.15, 4.16, 4.17, 4.18 and 4.19.

4.6 Stress Strain 3D Surface

The stress–strain curve of the whole stress field (p, q) can be obtained by drawing together the shear strain $\overline{\varepsilon}$, mean normal stress p and generalized shear stress q under two different water contents: $w_a = 26\%$, $w_b = 21.2\%$, as shown in Figs. 4.20, 4.21, 4.22 and 4.23. The three-dimensional surface graph is more vivid and intuitive

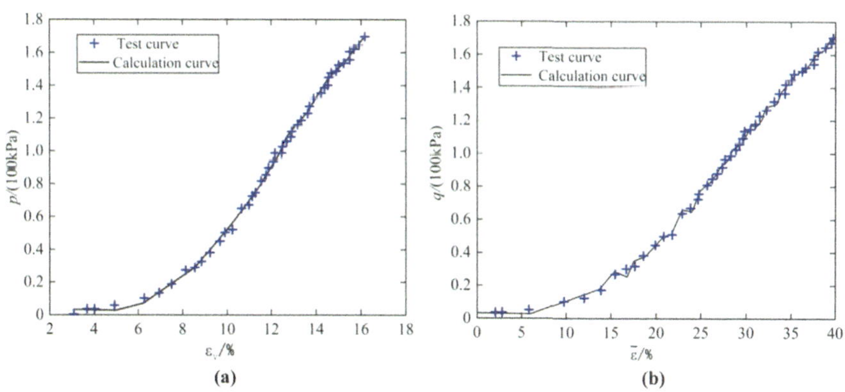

Fig. 4.14 Comparison of predicted and experimental values of sample B1-4

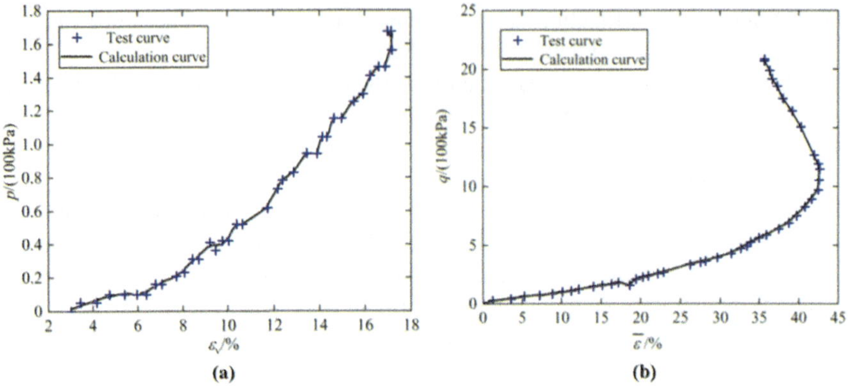

Fig. 4.15 Comparison of predicted and values experimental KT1

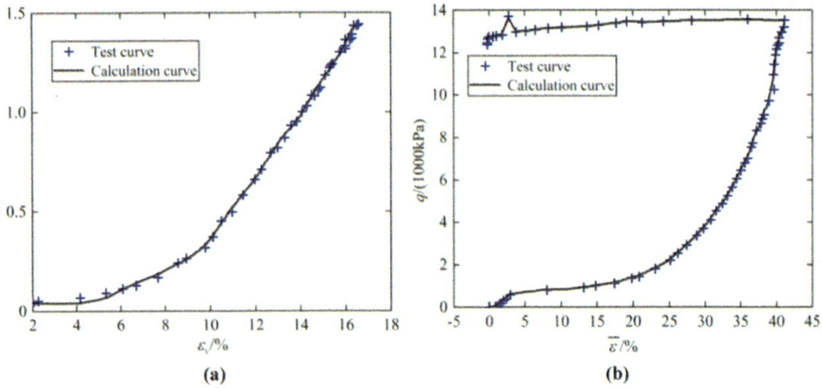

Fig. 4.16 Comparison of predicted and experimental values of KB1

Fig. 4.17 Comparison of predicted and experimental values of sample B1-1

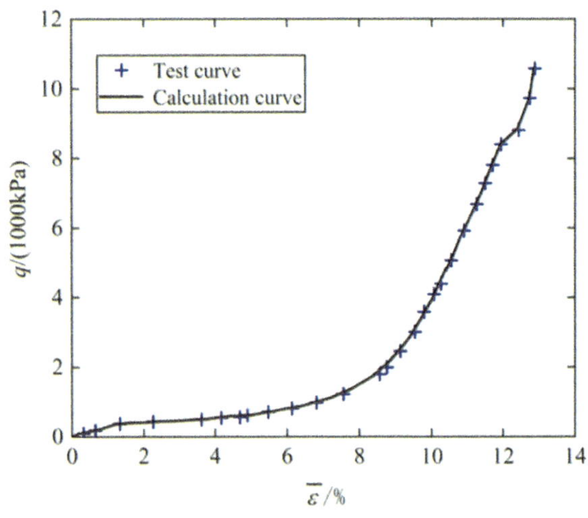

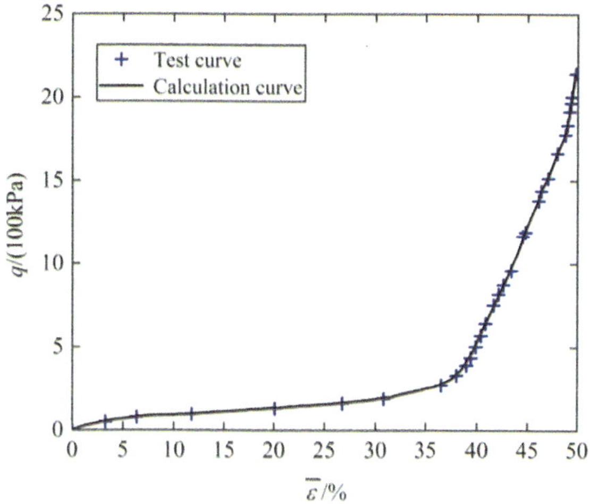

Fig. 4.18 Comparison of predicted and experimental values of sample B1-6

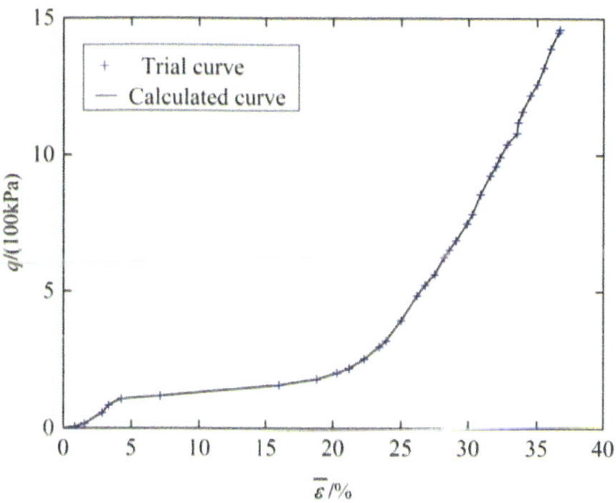

Fig. 4.19 Comparison of predicted value and test value of sample Bmix

than the two-dimensional graph, and it can comprehensively express the constitutive relation of soil. From the comparison of the stress–strain curves under the two water contents, it can be seen that the volume strain surface with water content of $w_a = 26\%$ is steeper and smoother than that of the sample with $w_b = 21.2\%$; The shear strain surface with water content $w_a = 26\%$ is more gentle than that with $w_b = 21.2\%$, and has protrusion in the middle. These differences show that the influence of water content on the stress–strain relationship is obvious.

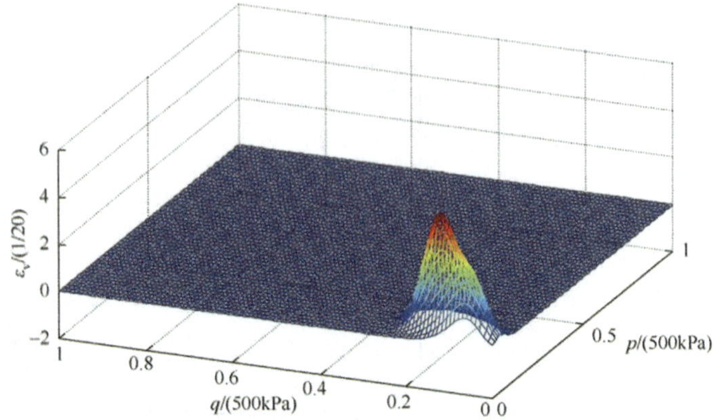

Fig. 4.20 Volume stress–strain surface of specimen a

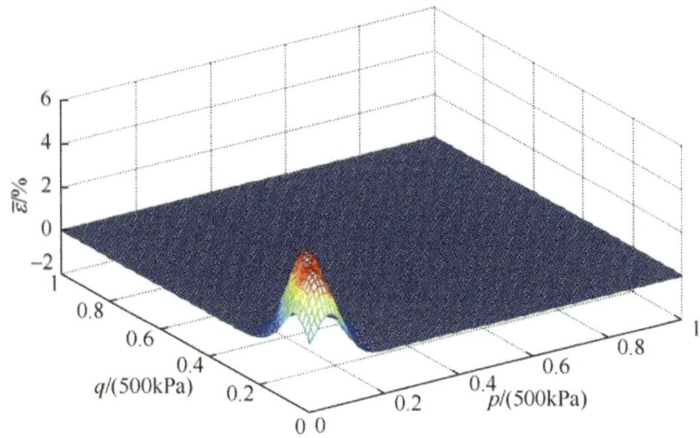

Fig. 4.21 Volume stress–strain surface of specimen b

Fig. 4.22 Shear stress–strain surface of specimen a

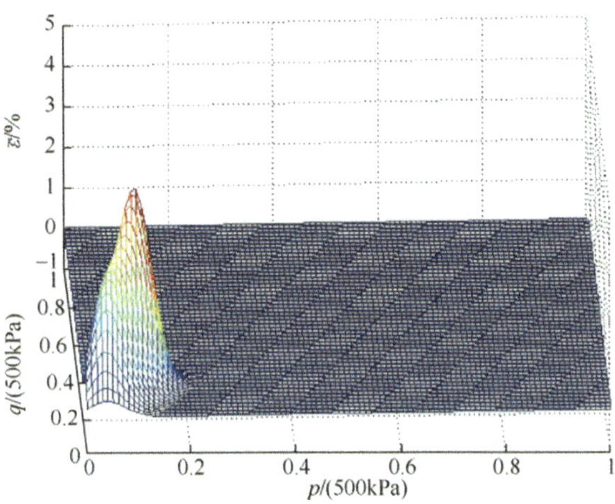

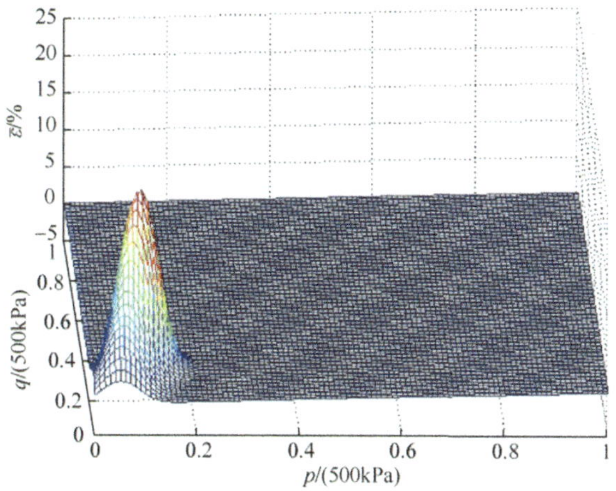

Fig. 4.23 Shear stress–strain surface of specimen b

4.7 Establishment of Numerical Model

The stress–strain relationship of soil is expressed by the following two functional functions:

$$\bar{\varepsilon} = F_2(p, q, p(q)) \tag{4.1}$$

$$\varepsilon_v = F_1(p, q, p(q)) \tag{4.2}$$

The stress–strain relationship reflects the coupling relationship between volume strain and shear strain and the cross effect of p and q, it not only considers the dilatancy and contraction of soil, but also reflects the influence of stress path on the constitutive relationship. Since the model parameters in this book are obtained by experimental simulation based on the actual stress path of soil, the influence of stress path on the constitutive relationship has been considered, so the above equations can be simplified as follows:

$$\varepsilon_{\mathrm{v}} = f_1(p, q) \tag{4.3}$$

$$\bar{\varepsilon} = f_2(p, q) \tag{4.4}$$

In the form of increment:

$$\mathrm{d}\varepsilon_{\mathrm{v}} = \frac{\partial f_1}{\partial p} dp + \frac{\partial f_1}{\partial q} dq \tag{4.5}$$

$$d\bar{\varepsilon} = \frac{\partial f_2}{\partial p} dp + \frac{\partial f_2}{\partial p} dq \tag{4.6}$$

The results of simultaneous Eqs. (4.5) and (4.6) are obtained:

$$dp = F d\varepsilon_v + G d\bar{\varepsilon} \tag{4.7}$$

$$dq = H d\varepsilon_v + K d\bar{\varepsilon} \tag{4.8}$$

where,

$$F = D/(AD - CB), G = -B/(AD - CB) \tag{4.9}$$

$$H = -C/(AD - CB), K = A/(AD - CB) \tag{4.10}$$

$$A = \frac{\partial f_1}{\partial p}, B = \frac{\partial f_1}{\partial q} \tag{4.11}$$

$$C = \frac{\partial f_2}{\partial p}, D = \frac{\partial f_2}{\partial q} \tag{4.12}$$

According to the theory of elasticity, the relationship between the principal stress $\sigma_1, \sigma_2, \sigma_3$ and p, q and Rhode angle θ is:

$$\begin{bmatrix} o_1 \\ \sigma_2 \\ \sigma_3 \end{bmatrix} = \begin{bmatrix} p \\ p \\ p \end{bmatrix} + \frac{2}{3} q \begin{bmatrix} \cos\theta \\ \cos(\theta - 2\pi/3) \\ \cos(\theta + 2\pi/3) \end{bmatrix} \tag{4.13}$$

where, $\sigma_1 \geq \sigma_2 \geq \sigma_3$. The influence of Rhode angle is not considered here, so if $\theta = 0$, then Eq. (4.13) becomes:

$$\begin{bmatrix} \sigma_1 \\ \sigma_2 \\ \sigma_3 \end{bmatrix} = \begin{bmatrix} p + \frac{2}{3} q \\ p - \frac{1}{3} q \\ p - \frac{1}{3} q \end{bmatrix} \tag{4.14}$$

The triaxial specimen in this book can be simplified as an axisymmetric problem because the shape and load of the object are symmetrical to the z-axis, so $\tau_{\theta rz} = \tau_{\theta z} = 0$. There are four stress components:σ_r, σ_z, τ_{rz} and σ_θ. The corresponding four strain components are:

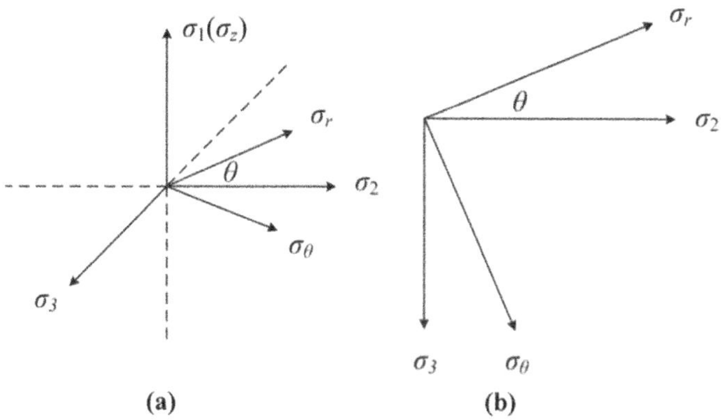

Fig. 4.24 Stress transformation of coordinate rotation

$$\varepsilon_r = \frac{\partial u}{\partial r}, \varepsilon_z = \frac{\partial w}{\partial z}, \varepsilon_\theta = \frac{u}{r}, \gamma = 2\varepsilon_{rz} = \frac{\partial u}{\partial z} + \frac{\partial w}{\partial r}$$

Let z-axis coincide with σ_1, then $\tau_{rz} = 0$. According to the stress transformation equation of coordinate axis rotation in elasticity, the relationship between principal stress σ_1, σ_2, σ_3 and stress σ_r, σ_z and σ_θ can be obtained as follows (the coordinate transformation is shown in Fig. 4.24):

$$\sigma_r = \sigma_2 \sin^2\theta + \sigma_3 \cos^2\theta \tag{4.15}$$

$$\sigma_\theta = \sigma_2 \cos^2\theta + \sigma_3 \sin^2\theta \tag{4.16}$$

Due to the limitation of experimental conditions, the conventional triaxial test of expansive soil is carried out in this book. Therefore, $\sigma_2 = \sigma_3$, so we can get:

$$\sigma 2 = \sigma_3 = \sigma_r = \sigma 0 \tag{4.17}$$

Combining with Eqs. (4.14) and (4.17), the relationship between stress σ_r, σ_z, σ_θ and p, q in axisymmetric problem can be obtained as follows:

$$\begin{bmatrix} \mathrm{d}\sigma_z \\ \mathrm{d}\sigma_r \\ \mathrm{d}\sigma_\theta \end{bmatrix} = \begin{bmatrix} \mathrm{d}p + \frac{2}{3}\mathrm{d}q \\ \mathrm{d}p - \frac{1}{3}\mathrm{d}q \\ \mathrm{d}p - \frac{1}{3}\mathrm{d}q \end{bmatrix} \tag{4.18}$$

For axisymmetric cases, there are also:

$$\varepsilon_v = \varepsilon_r + \varepsilon_z + \varepsilon_\theta \tag{4.19}$$

$$\overline{\varepsilon} = \varepsilon_a - \frac{1}{3}\varepsilon_v = \varepsilon_z - \frac{1}{3}(\varepsilon_r + \varepsilon_z + \varepsilon_\theta) \tag{4.20}$$

The above two types of differentiation can be obtained:

$$d\varepsilon_v = d\varepsilon_r + d\varepsilon_\theta + d\varepsilon_z \tag{4.21}$$

$$d\overline{\varepsilon} = -\frac{1}{3}d\varepsilon_r + \frac{2}{3}d\varepsilon_z - \frac{1}{3}d\varepsilon_\theta \tag{4.22}$$

Combining Eqs. (4.7), (4.8), (4.19), (4.21) and (4.22) can obtain the elastic–plastic stress–strain relationship of axisymmetric sand samples:

$$\begin{bmatrix} d\sigma_z \\ d\sigma_r \\ d\sigma_\theta \end{bmatrix} = \begin{bmatrix} U_{zz} & U_{zr} & U_{z\theta} \\ U_{rz} & U_{rr} & U_{r\theta} \\ U_{\theta z} & U_{\theta r} & U_{\theta\theta} \end{bmatrix} \begin{bmatrix} d\varepsilon_z \\ d\varepsilon_r \\ d\varepsilon_\theta \end{bmatrix} = \left[D_{ep}\right] \begin{bmatrix} d\varepsilon_z \\ d\varepsilon_r \\ d\varepsilon_\theta \end{bmatrix} \tag{4.23}$$

where, the elements in the elastoplastic stiffness matrix D_{ep} are

$$\begin{aligned}
U_{zz} &= F + \frac{2}{3}H + \frac{2}{3}G + \frac{4}{9}K \\
U_{zr} &= F + \frac{2}{3}H - \frac{1}{3}G - \frac{2}{9}K \\
U_{z\theta} &= F + \frac{2}{3}H - \frac{1}{3}G - \frac{2}{9}K \\
U_{rz} &= F - \frac{1}{3}H + \frac{2}{3}G - \frac{2}{9}K \\
U_{rr} &= F - \frac{1}{3}H - \frac{1}{3}G + \frac{1}{9}K \\
U_{r\theta} &= F - \frac{1}{3}H - \frac{1}{3}G + \frac{1}{9}K \\
U_{\theta z} &= F - \frac{1}{3}H + \frac{2}{3}G - \frac{2}{9}K \\
U_{\theta r} &= F - \frac{1}{3}H - \frac{1}{3}G + \frac{1}{9}K \\
U_{\theta\theta} &= F - \frac{1}{3}H - \frac{1}{3}G + \frac{1}{9}K
\end{aligned} \tag{4.24}$$

Here, the average normal stress p and the generalized shear stress q are used as the input parameters of the designed neural network, taking volume strain ε_v and shear strain $\overline{\varepsilon}$ as the output parameters of the network, the mapping effect of functions f_1 and f_2 will be replaced by a trained network. After the network training is mature, the parameters A, B, C, and D in the above matrix can be obtained by the prediction function of the network, and the elastoplastic stiffness matrix, that is, the numerical model of the elastoplastic constitutive relationship of expansive soil can be obtained.

4.8 Verification of the Numerical Model

In order to verify the effectiveness and superiority of the neural network-based numerical modeling method, this section incorporates the established neural network constitutive model into the finite element program, analyze the triaxial specimen used in the test, and compare the calculated stress–strain relationship curve with the actual relationship curve in the test.

According to the derivation in Sect. 4.7, for a trained network, two constitutive equations are obtained by central difference method four coefficients of functional $\partial f_1(p,q)/\partial p$, $\partial f_1(p,q)/\partial q$, $\partial f_2(p,q)/\partial p$, $\partial f_2(p,q)/\partial q$, specific steps are as follows:

(1) Taking the given p_1 and q_1 as the input of the network, substituting it into the network can obtain the output y_1 and y_2;
(2) Using $p-\Delta x, q_1$ as the input of the network, substituting them into the network can obtain output y_1'', y_2'';
(3) Using p_1, $q_1 - \Delta x$ as the input of the network, substituting them into the network can obtain output y_1'', y_2'';
(4) Using $p+\Delta x, q_1$ as the input of the network, substituting them into the network can obtain output y_1^1, y_2^1;
(5) Using p_1, $q_1 + \Delta x$ as the input of the network, substituting them into the network can obtain output y_1^2, y_2^2;

Then find the value of $\partial f_1/\partial p$, $\partial f_1/\partial q$, $\partial f_2/\partial p$, $\partial f_2/\partial q$ according to the following formulas:

$$\left(\frac{\partial f_1(p,q)}{\partial p}\right)_1 = \frac{y_1 - y_1''}{\Delta x} \tag{4.25}$$

$$\left(\frac{\partial f_2(p,q)}{\partial p}\right)_1 = \frac{y_2 - y_2''}{\Delta x} \tag{4.26}$$

$$\left(\frac{\partial f_1(p,q)}{\partial q}\right)_1 = \frac{y_1 - y_1''}{\Delta x} \tag{4.27}$$

$$\left(\frac{\partial f_2(p,q)}{\partial q}\right)_1 = \frac{y_2 - y_2''}{\Delta x} \tag{4.28}$$

$$\left(\frac{\partial f_1(p,q)}{\partial p}\right)_2 = \frac{y_1^1 - y_1}{\Delta x} \tag{4.29}$$

$$\left(\frac{\partial f_2(p,q)}{\partial p}\right)_2 = \frac{y_2^1 - y_2}{\Delta x} \tag{4.30}$$

$$\left(\frac{\partial f_1(p,q)}{\partial q}\right)_2 = \frac{y_1^2 - y_1}{\Delta x} \tag{4.31}$$

$$\left(\frac{\partial f_2(p,q)}{\partial q}\right)_2 = \frac{y_2^2 - y_2}{\Delta x} \tag{4.32}$$

$$\left(\frac{\partial f_1(p,q)}{\partial p}\right) = \frac{1}{2}\left[\left(\frac{\partial f_1(p,q)}{\partial p}\right)_1 + \left(\frac{\partial f_1(p,q)}{\partial p}\right)_2\right] \tag{4.33}$$

$$\left(\frac{\partial f_1(p,q)}{\partial p}\right) = \frac{1}{2}\left[\left(\frac{\partial f_1(p,q)}{\partial p}\right)_1 + \left(\frac{\partial f_1(p,q)}{\partial q}\right)_2\right] \tag{4.34}$$

$$\left(\frac{\partial f_2(p,q)}{\partial p}\right)_2 = \frac{1}{2}\left[\left(\frac{\partial f_2(p,q)}{\partial p}\right)_1 + \left(\frac{\partial f_2(p,q)}{\partial p}\right)_2\right] \tag{4.35}$$

$$\left(\frac{\partial f_2(p,q)}{\partial q}\right)_2 = \frac{1}{2}\left[\left(\frac{\partial f_2(p,q)}{\partial q}\right)_1 + \left(\frac{\partial f_2(p,q)}{\partial q}\right)_2\right] \tag{4.36}$$

The size of Δx has a great influence on the derivative value. If it is too small, the mapping relationship reflected by the network is different from the true constitutive relationship. As Δx decreases, the difference becomes more and more significant, and the obtained derivative value will have a larger error; if Δx is too large, the approximation ability of the network will not be fully applied, and large errors will also occur, the load step size x = 30 kPa is selected in this book.

After the above four coefficients are calculated, they are applied to the elastic–plastic matrix $\left[D_{ep}\right]$ derived in Sect. 4.7.

4.8.1 Calculation Model

The test data of triaxial specimen B1-4 under confining pressure of 50 and 300 kPa along the path of triaxial compression (CTC) were selected, use this as a training sample to establish a corresponding constitutive model. Substituting the constitutive model into the finite element, the confining pressure of the triaxial specimen along the CTC path is analyzed as Stress-strain relationship under 100 kPa. The diameter of triaxial specimen is 3.91 cm, because the specimen shape and load are axisymmetric, the nodes on the symmetry axis are only vertical Deformation of direction. The meshing of the finite element of the calculation model is shown in Fig. 4.25.

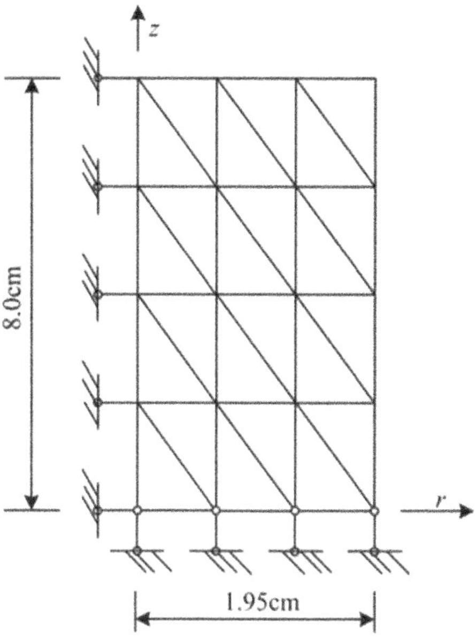

Fig. 4.25 Finite element meshing of triaxial

4.8.2 Calculation Results

Take the calculated value of the sample center point as the calculation result, compare the calculated $\varepsilon_v - p$ and $\overline{\varepsilon} - q$ curves with the test curve, the result is shown in Fig. 4.26. The solid line in the Fig. 4.26 is the test curve, and the dashed line is the calculation curve. It can be seen from the Fig. 4.26 that the calculated stress–strain curve is quite close to the actual curve in the test. It shows that the neural network

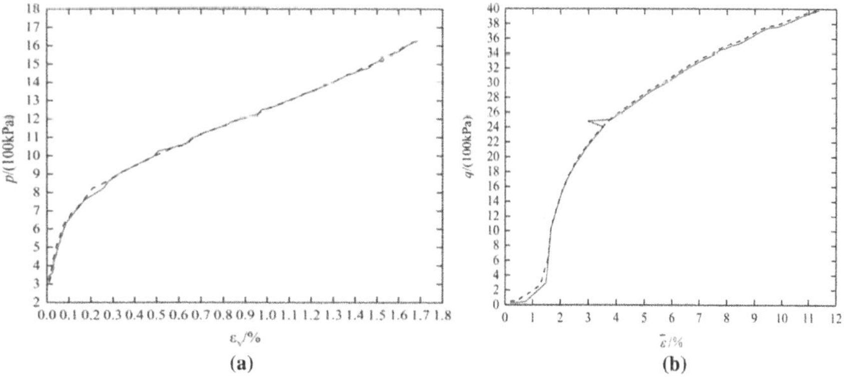

Fig. 4.26 Comparison of finite element calculation value and test value of sample

numerical model of expansive soil can better reflect the constitutive relationship of the sample.

4.8.3 Conclusion

Using neural network as a tool, a numerical method for establishing a rock-soil constitutive model is proposed, and the feasibility and effectiveness of the model are verified through case analysis. The results show that the numerical modeling method of rock-soil constitutive relationship has the following advantages:

(1) The stress–strain relationship is directly extracted from the test curve and does not involve plastic potential theory. Overcome the difficulty of finding the analytical expression of the plastic potential;

(2) It can reflect the influence of dilatancy, shrinkage and initial conditions, it provides an effective method for simulating the actual stress path in the soil.

Chapter 5
Soil Triaxial Test and Numerical Modeling

In this chapter, the triaxial compression test of sand is carried out stress and strain test data under p path. Then, according to the theory of numerical modeling, the numerical model of sand under the path of equal principal stress ratio is established. Finally, the model is verified.

5.1 Sandy Soil Physical Index Determination Test

Before triaxial test of sand, it is necessary to know the physical indexes of sand, such as density and particle size distribution. The following contents were carried out particle screening test and density measurement test.

5.1.1 Sand Particle Analysis Test

The test sand is taken from the construction site of high-rise residential buildings in the West District of Huazhong University of science and technology, which does not meet the test requirements. Therefore, the original sand is re graded. The result is medium dense sand.

(1) Drying sample 1000 g is called pick.
(2) The sample is poured into a standard sieve stacked in turn. Then the sieve is placed on the sieve machine for a sieve time 10–15 min.
(3) Starting with a sieve with the largest aperture, each sieve is removed in sequence. The samples are weighed on the sieve at all levels and inside the chassis quality, accurate to 0.1 g.
(4) The difference between the sum of the quality of the sieve and the sand at the bottom of the sieve should not be greater than 1% with the quality of the pre-screening sample.

Q. Ren and J. Zhou, *Numerical Modeling of Soil Constitutive Relationship*,
https://doi.org/10.1007/978-981-16-3231-0_5

Table 5.1 Particle size analysis of sand

Mesh size (mm)	Residue (g)	Screening allowance (%)	Accumulated screen residue (%)
5	0	0	0
2.5	54.73	5.48	5.48
1.25	152.52	15.27	20.75
0.63	351.68	35.21	55.96
Below 0.63	439.87	44.04	100

Note The trial date is April 5, 2002

The standard sieve diameters used in the test are 5, 2.5, 1.25, 0.63 mm and below, and the results are in Table 5.1.

5.1.2 *Surface Density Determination of Sand*

The appearance density is commonly known as the weight, which refers to the mass of the material per unit volume in its natural state. Due to space, the test steps are omitted. The results of the experiment are in Table 5.2.

The formula for calculating the surface density is

$$\rho_0 = \frac{G_0}{G_0 + G_2 - G_1} \tag{5.1}$$

Taking the average twice and the surface density is

$$\rho_0 = \frac{2.646 + 2.662}{2} = 2.654\ (\text{g/cm}^3) \tag{5.2}$$

Table 5.2 The test table for the determination of the surface density of sand and soil

Number of trials	Dry sand quality G_0 (g)	Dry sand + water + bottle quality G_1 (g)	Refill the water + bottle quality G_2 (g)	The surface density ρ_0 (g/cm^3)
1	300	853.6	667.0	2.646
2	300	858.9	671.6	2.662

Note The trial date is April 8, 2002

Table 5.3 Sand accumulation density determination test table

Number of trials	Capacity barrel G_I (g)	Sand + capacity bottle $'_2$ (g)	Stacked density ρ'_0 (g/cm^3)
1	405.3	1858.5	1.453
2	405.3	1867.4	1.462

Note The trial date is April 10, 2002

5.1.3 Sand Accumulation Density

Stacking density refers to the unit volume mass of powdered or granular material in the accumulation state. The results of the experiment can be found in Table 5.3.

The calculation formula for stacking density is

$$\rho_0' = \frac{G_2{-}G_1}{1000} \tag{5.3}$$

Take the average twice, and the stacking density is

$$\rho_0' = \frac{1.453 + 1.462}{2} = 1.458\ (\text{g/cm}^3) \tag{5.4}$$

5.2 Sand Three-Axis Test

5.2.1 Testscheme

Three kinds of stress path triaxial tests are carried out on medium dense sand. The details are as follows: ① Path 1 is equal principal stress ratio path (ERTC). Cyclic loading tests with $k = \sigma_3/\sigma_1 = 0.3, 0.42, 0.5, 0.6$ and 0.7 are carried out. ② Path 2 is the conventional compression test (DCTC). Under the condition of isotropic consolidation, the cyclic loading tests of three confining pressures of $\sigma_3 = 100$ kPa, 200 kPa, 300 kPa, 400 kPa and 500 kPa are carried out. ③ Path 3 is equal p test (DPTC). Cyclic loading test with average normal stress $p = 100$ kPa, 200 kPa, 300 kPa, 400 kPa, 500 kPa are carried out, under isotropic consolidation condition. The three stress paths are shown in Fig. 5.1.

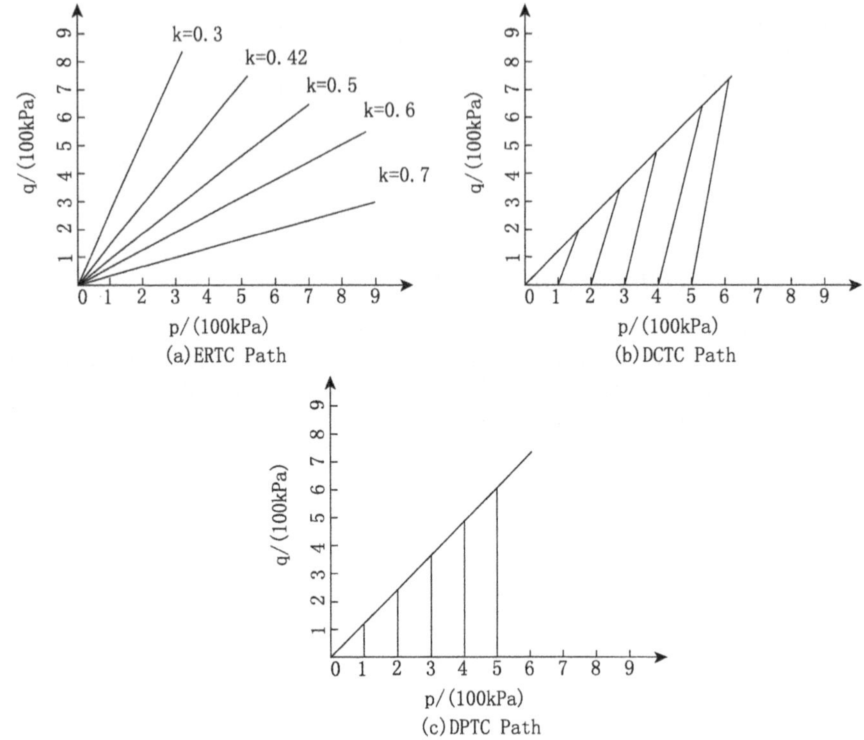

Fig. 5.1 Three stress path maps

5.2.2 Sand Sample Preparation

Since the test requires the use of medium dense sand, according to the provisions of the code for geotechnical engineering investigation, the porosity ratio *e* in natural state is taken as the standard for classifying the compactness of sand, as shown in Table 5.4.

Assuming that the sand sample weighs 600 g, the sample has a diameter of 6.18 cm and a height of 12.5 cm, so the calculation volume formula for the sample is

Table 5.4 Sand accumulation density determination test table

Soil	Density degree			
	density	Middle density	Slow density	Loose
Gravel, coarse sand, medium sand	$e < 0.60$	$0.60 \le e \le 0.75$	$0.75 \le e \le 0.85$	$e > 0.85$
Fine sand, powdered sand	$e < 0.70$	$0.70 \le e \le 0.85$	$0.85 \le e \le 0.95$	$e > 0.95$

Table 5.5 Preparation of sand sample

Particle diameter rang (mm)	Proportion (%)	Quality (%)
2.50–5.0	5.48	32.9
1.25–2.50	15.27	91.6
0.63–1.25	35.21	211.3
0–0.63	44.04	264.2
Sum	100	600

$$V = \frac{\pi \times 6.18^2}{4} \times 12.5 \approx 374.95\,(\text{cm}^3) \tag{5.5}$$

where the solid volume is

$$V_s = \frac{m_s}{\rho_0} = \frac{600}{2.654} \approx 226.07\,(\text{cm}^3) \tag{5.6}$$

The porosity ratio is

$$e = \frac{V - V_s}{V_s} = \frac{374.95 - 226.07}{226.07} \approx 0.66 \tag{5.7}$$

The porosity ratio of 0.66 meets the pore ratio range of medium-tight sand in Table 5.4, so the availage of each sand is seen in Table 5.5.

5.3 Test Data Processing and Analysis

5.3.1 Three-Axis Test Data

Tables 5.6 and 5.7 show the test data under the path of equal principal stress ratio. Due to space limitation, only the sample data for $k = 0.42$ and 0.6 are listed.

5.3.2 Determination of Hydrostatic Test Data and K *Value*

(1) In order to determine the elastic bulk modulus K, it is necessary to carry out the isotropic consolidation test [1]. The compression test of soil under hydrostatic pressure $p = \sigma_1 = \sigma_2 = \sigma_3$ is carried out. Table 5.8 shows the hydrostatic test data.

(2) Determination of elastic bulk modulus K

Table 5.6 Triaxial test data with equal principal stress ratio $k = 0.42$

Axial deformation (μm)	Measure the force ring (mm)	Drain tube reading (mm)	Axial strain (%)	Correct the area (mm^2)	Main stress difference (kPa)	surrounding pressure (kPa)	water discharge (m)	Major principal stress (kPa)	P (kPa)	Q (kPa)	Volume strain (%)	Shear strain (%)	K	Plastic volume deformation (%)	1/G (mm/N)	Plastic shear (%)
298	2.5	19.5	2.55	30.78	0.21	0.157	0.8	0.36	0.23	0.21	0.23	2.47	217.02	0.23	0.013158	2.47
370	5	18.6	3.16	30.98	0.41	0.314	1.7	0.73	0.45	0.41	0.48	3.00	338.64	0.48	0.013158	3.00
381	7.5	17.4	3.26	31.01	0.62	0.469	2.9	1.09	0.67	0.62	0.83	2.98	438.71	0.82	0.013158	2.97
396	10	16.7	3.38	31.05	0.82	0.625	3.6	1.45	0.90	0.82	1.03	3.04	527.70	1.02	0.013158	3.03
408	12.5	16.1	3.49	31.08	1.03	0.781	4.2	1.81	1.12	1.03	1.20	3.09	608.99	1.19	0.013158	3.07
417	15	15.5	3.56	31.11	1.23	0.938	4.8	2.17	1.35	1.23	1.37	3.11	684.98	1.37	0.013158	3.09
425	17.5	15.1	3.63	31.13	1.43	1.094	5.2	2.53	1.57	1.43	1.48	3.14	756.23	1.48	0.013158	3.12
434	20	14.6	3.71	31.16	1.64	1.250	5.7	2.89	1.80	1.64	1.62	3.17	823.89	1.62	0.013158	3.15
441	22.5	14	3.77	31.18	1.84	1.406	6.3	3.25	2.02	1.84	1.80	3.17	888.61	1.79	0.013158	3.15
450	25	13.7	3.85	31.20	2.04	1.563	6.6	3.61	2.24	2.04	1.88	3.22	951.03	1.88	0.013158	3.19
454	27.5	13.5	3.88	31.21	2.25	1.719	6.8	3.97	2.47	2.25	1.94	3.23	1011.05	1.94	0.013158	3.20
460	30	13.2	3.93	31.23	2.45	1.875	7.1	4.32	2.69	2.45	2.02	3.26	1069.11	2.02	0.013158	3.23
467	32.5	12.9	3.99	31.25	2.65	2.031	7.4	4.68	2.92	2.65	2.11	3.29	1125.44	2.11	0.013158	3.25
474	35	12.8	4.05	31.27	2.85	2.188	7.5	5.04	3.14	2.85	2.14	3.34	1180.47	2.13	0.013158	3.30
480	37.5	12.5	4.10	31.28	3.06	2.344	7.8	5.40	3.36	3.06	2.22	3.36	1233.89	2.22	0.013158	3.32
487	40	12.3	4.16	31.30	3.26	2.500	8	5.76	3.59	3.26	2.28	3.40	1286.01	2.28	0.013158	3.36
492	42.5	12.2	4.21	31.32	3.46	2.656	8.1	6.12	3.81	3.46	2.31	3.44	1337.01	2.31	0.013158	3.39
499	45	12	4.26	31.34	3.66	2.813	8.3	6.47	4.03	3.66	2.36	3.48	1387.12	2.36	0.013158	3.43
506	47.5	11.7	4.32	31.36	3.86	2.969	8.6	6.83	4.26	3.86	2.45	3.51	1436.01	2.45	0.013158	3.46
506	45	11.6	4.32	31.36	3.66	2.813	8.7	6.47	4.03	3.66	2.48	3.50	1386.95	2.48	0.013158	3.45
504	40	11.8	4.31	31.35	3.25	2.500	8.5	5.75	3.58	3.25	2.42	3.50	1285.63	2.42	0.013158	3.46
501.5	35	12	4.29	31.34	2.85	2.188	8.3	5.04	3.14	2.85	2.36	3.50	1179.91	2.36	0.013158	3.46

(continued)

Table 5.6 (continued)

Axial deformation (μm)	Measure the force ring (mm)	Drain tube reading (mm)	Axial strain (%)	Correct the area (mm^2)	Main stress difference (kPa)	surrounding pressure (kPa)	water discharge (m)	Major principal stress (kPa)	P (kPa)	Q (kPa)	Volume strain (%)	Shear strain (%)	K	Plastic volume deformation (%)	1/G (mm/N)	Plastic shear (%)
498.5	30	12.2	4.26	31.34	2.44	1.875	8.1	4.32	2.69	2.44	2.31	3.49	1068.40	2.31	0.013158	3.46
495	25	12.5	4.23	31.33	2.04	1.563	7.8	3.60	2.24	2.04	2.22	3.49	950.28	2.22	0.013158	3.46
491	20	12.7	4.20	31.31	1.63	1.250	7.6	2.88	1.79	1.63	2.17	3.47	823.08	2.16	0.013158	3.45
487	15	13	4.16	31.30	1.22	0.938	7.3	2.16	1.35	1.22	2.08	3.47	684.15	2.08	0.013158	3.45
481.5	10	13.5	4.12	31.29	0.82	0.625	6.8	1.44	0.90	0.82	1.94	3.47	526.91	1.94	0.013158	3.46
478	5	14.5	4.09	31.28	0.41	0.314	5.8	0.72	0.45	0.41	1.65	3.53	338.01	1.65	0.013158	3.53
488	10	13.8	4.17	31.31	0.81	0.625	6.5	1.44	0.90	0.81	1.85	3.55	526.85	1.85	0.013158	3.54
493	15	13	4.21	31.32	1.22	0.938	7.3	2.16	1.35	1.22	2.08	3.52	684.08	2.08	0.013158	3.50
499.5	20	12.7	4.27	31.34	1.63	1.250	7.6	2.88	1.79	1.63	2.17	3.55	822.95	2.16	0.013158	3.53
504	25	12.5	4.31	31.35	2.03	1.563	7.8	3.60	2.24	2.03	2.22	3.57	950.13	2.22	0.013158	3.54
509	30	12.2	4.35	31.36	2.44	1.875	8.1	4.31	2.69	2.44	2.31	3.58	1068.20	2.31	0.013158	3.55
514	35	12	4.39	31.38	2.84	2.188	8.3	5.03	3.14	2.84	2.36	3.60	1179.65	2.36	0.013158	3.57
520	40	11.7	4.44	31.40	3.25	2.500	8.6	5.75	3.58	3.25	2.45	3.63	1285.27	2.45	0.013158	3.58
526	45	11.5	4.50	31.41	3.65	2.813	8.8	6.47	4.03	3.65	2.51	3.66	1386.47	2.50	0.013158	3.61
531	47.5	11.3	4.54	31.43	3.85	2.969	9	6.82	4.25	3.85	2.56	3.68	1435.39	2.56	0.013158	3.63
534	50	11	4.56	31.43	4.06	3.125	9.3	7.18	4.48	4.06	2.65	3.68	1483.44	2.65	0.013158	3.63
537	52.5	11	4.59	31.44	4.26	3.281	9.3	7.54	4.70	4.26	2.65	3.71	1530.63	2.65	0.013158	3.65
542	55	10.8	4.63	31.46	4.46	3.438	9.5	7.90	4.92	4.46	2.71	3.73	1577.18	2.70	0.013158	3.67
547	57.5	10.6	4.68	31.47	4.66	3.594	9.7	8.25	5.15	4.66	2.76	3.75	1622.77	2.76	0.013158	3.69
556	60	10.4	4.75	31.50	4.86	3.750	9.9	8.61	5.37	4.86	2.82	3.81	1667.52	2.82	0.013158	3.75
557	62.5	10.2	4.76	31.50	5.06	3.906	10.1	8.97	5.59	5.06	2.88	3.80	1711.84	2.87	0.013158	3.73
561	65	10.1	4.79	31.51	5.26	4.063	10.2	9.32	5.82	5.26	2.91	3.83	1755.64	2.90	0.013158	3.76

(continued)

Table 5.6 (continued)

Axial deformation (μm)	Measure the force ring (mm)	Drain tube reading (mm)	Axial strain (%)	Correct the area (mm^2)	Main stress difference (kPa)	surrounding pressure (kPa)	water discharge (m)	Major principal stress (kPa)	*P* (kPa)	*Q* (kPa)	Volume strain (%)	Shear strain (%)	*K*	Plastic volume deformation (%)	1/G (mm/N)	Plastic shear (%)
569	67.5	9.8	4.86	31.53	5.46	4.219	10.5	9.68	6.04	5.46	2.99	3.87	1798.51	2.99	0.013158	3.79
573	70	9.7	4.90	31.54	5.66	4.375	10.6	10.03	6.26	5.66	3.02	3.89	1840.93	3.02	0.013158	3.82
580	72.5	9.4	4.96	31.56	5.86	4.531	10.9	10.39	6.48	5.86	3.11	3.92	1882.71	3.10	0.007321	3.88
586	75	9.3	5.01	31.58	6.06	4.688	11	10.74	6.71	6.06	3.13	3.96	1924.18	3.13	0.007321	3.92
593	77.5	9.1	5.07	31.60	6.25	4.844	11.2	11.10	6.93	6.25	3.19	4.00	1964.94	3.19	0.007321	3.96
596	80	9	5.09	31.61	6.45	5.033	11.3	11.49	7.18	6.45	3.22	4.02	2011.31	3.22	0.007321	3.97
599	82.5	8.8	5.12	31.62	6.65	5.156	11.5	11.81	7.37	6.65	3.28	4.03	2045.32	3.27	0.007321	3.98
603	85	8.7	5.15	31.63	6.85	5.313	11.6	12.17	7.60	6.85	3.31	4.05	2085.00	3.30	0.007321	4.00
608	87.5	8.6	5.20	31.64	7.05	5.469	11.7	12.52	7.82	7.05	3.33	4.09	2124.04	3.33	0.007321	4.03
615	90	8.4	5.26	31.66	7.25	5.625	11.9	12.87	8.04	7.25	3.39	4.13	2162.59	3.39	0.007321	4.07
620	92.5	8	5.30	31.68	7.45	5.781	12.3	13.23	8.26	7.45	3.50	4.13	2200.84	3.50	0.007321	4.08
630	95	7.8	5.38	31.71	7.64	5.938	12.5	13.58	8.48	7.64	3.56	4.20	2238.68	3.56	0.007321	4.14
630	90	7.8	5.38	31.71	7.24	5.625	12.5	12.86	8.04	7.24	3.56	4.20	2162.03	3.56	0.007321	4.14
628	85	7.9	5.37	31.70	6.84	5.313	12.4	12.15	7.59	6.84	3.53	4.19	2084.09	3.53	0.007321	4.14
626	80	8	5.35	31.70	6.44	5.033	12.3	11.47	7.18	6.44	3.50	4.18	2010.26	3.50	0.007321	4.14
62 4	75	8.2	5.33	31.69	6.04	4.688	12.1	10.72	6.70	6.04	3.45	4.18	1922.91	3.44	0.007321	4.14
622	70	8.3	5.32	31.68	5.63	4.375	12	10.01	6.25	5.63	3.42	4.18	1839.36	3.42	0.007321	4.14
620	65	8.4	5.30	31.68	5.23	4.063	11.9	9.30	5.81	5.23	3.39	4.17	1753.83	3.39	0.007321	4.13
618	60	8.5	5.28	31.67	4.83	3.750	11.8	8.58	5.36	4.83	3.36	4.16	1665.72	3.36	0.007321	4.13
615	55	8.7	5.26	31.66	4.43	3.438	11.6	7.87	4.91	4.43	3.31	4.15	1575.18	3.30	0.007321	4.12
613	50	8.8	5.24	31.66	4.03	3.125	11.5	7.15	4.47	4.03	3.28	4.15	1481.40	3.27	0.007321	4.12
610	45	8.9	5.21	31.65	3.63	2.813	11.4	6.44	4.02	3.63	3.25	4.13	1384.44	3.25	0.007321	4.10

(continued)

Table 5.6 (continued)

Axial deformation (μm)	Measure the force ring (mm)	Drain tube reading (mm)	Axial strain (%)	Correct the area (mm^2)	Main stress difference (kPa)	surrounding pressure (kPa)	water discharge (m)	Major principal stress (kPa)	P (kPa)	Q (kPa)	Volume strain (%)	Shear strain (%)	K	Plastic volume deformation (%)	1/G (mm/N)	Plastic shear (%)
608	40	9.1	5.20	31.64	3.22	2.500	11.2	5.72	3.57	3.22	3.19	4.13	1283.30	3.19	0.007321	4.11
605	35	9.3	5.17	31.64	2.82	2.188	11	5.01	3.13	2.82	3.13	4.13	1177.79	3.13	0.007321	4.11
602	30	9.5	5.15	31.63	2.42	1.875	10.8	4.29	2.68	2.42	3.08	4.12	1066.47	3.07	0.007321	4.10
599	25	9.8	5.12	31.62	2.02	1.563	10.5	3.58	2.24	2.02	2.99	4.12	948.56	2.99	0.007321	4.11
595	20	10.2	5.09	31.61	1.61	1.250	10.1	2.86	1.79	1.61	2.88	4.13	821.59	2.88	0.007321	4.11
591	15	10.7	5.05	31.60	1.21	0.938	9.6	2.15	1.34	1.21	2.74	4.14	682.91	2.73	0.007321	4.13
586	10	11.3	5.01	31.58	0.81	0.625	9	1.43	0.89	0.81	2.56	4.15	525.96	2.56	0.007321	4.15
584	5	12.8	4.99	31.58	0.40	0.314	7.5	0.72	0.45	0.40	2.14	4.28	337.39	2.14	0.007321	4.28
599	10	12.1	5.12	31.62	0.81	0.625	8.2	1.43	0.89	0.81	2.34	4.34	525.84	2.33	0.007321	4.33
605	15	11.3	5.17	31.64	1.21	0.938	9	2.15	1.34	1.21	2.56	4.32	682.75	2.56	0.007321	4.31
609	20	11	5.21	31.65	1.61	1.250	9.3	2.86	1.79	1.61	2.65	4.32	821.39	2.65	0.007321	4.31
615	25	10.5	5.26	31.66	2.01	1.563	9.8	3.58	2.23	2.01	2.79	4.33	948.30	2.79	0.007321	4.31
619	30	10.1	5.29	31.68	2.42	1.875	10.2	4.29	2.68	2.42	2.91	4.32	1066.16	2.90	0.007321	4.30
623	35	9.9	5.32	31.69	2.82	2.188	10.4	5.00	3.13	2.82	2.96	4.34	1177.42	2.96	0.007321	4.32
627	40	9.6	5.36	31.70	3.22	2.500	10.7	5.72	3.57	3.22	3.05	4.34	1282.87	3.05	0.007321	4.32
631	45	9.4	5.39	31.71	3.62	2.813	10.9	6.43	4.02	3.62	3.11	4.36	1383.93	3.10	0.007321	4.33
636	50	9.2	5.44	31.72	4.02	3.125	11.1	7.14	4.46	4.02	3.16	4.38	1480.80	3.16	0.007321	4.35
639	55	9	5.46	31.73	4.42	3.438	11.3	7.86	4.91	4.42	3.22	4.39	1574.52	3.22	0.007321	4.36
643	60	8.7	5.50	31.74	4.82	3.750	11.6	8.57	5.36	4.82	3.31	4.39	1664.99	3.30	0.007321	4.36
647	65	8.5	5.53	31.76	5.22	4.063	11.8	9.28	5.80	5.22	3.36	4.41	1753.00	3.36	0.007321	4.37
652	70	8.3	5.57	31.77	5.62	4.375	12	9.99	6.25	5.62	3.42	4.43	1838.39	3.42	0.007321	4.39
656	75	8.2	5.61	31.78	6.02	4.688	12.1	10.71	6.69	6.02	3.45	4.46	1921.84	3.44	0.007321	4.41

(continued)

Table 5.6 (continued)

Axial deformation (μm)	Measure the force ring (mm)	Drain tube reading (mm)	Axial strain (%)	Correct the area (mm^2)	Main stress difference (kPa)	surrounding pressure (kPa)	water discharge (m)	Major principal stress (kPa)	*P* (kPa)	*Q* (kPa)	Volume strain (%)	Shear strain (%)	*K*	Plastic volume deformation (%)	1/G (mm/N)	Plastic shear (%)
660	80	7.8	5.64	31.79	6.42	5.033	12.5	11.45	7.17	6.42	3.56	4.45	2009.08	3.56	0.007321	4.41
664	85	7.7	5.68	31.81	6.81	5.313	12.6	12.13	7.58	6.81	3.59	4.48	2082.78	3.59	0.007321	4.43
668	90	7.6	5.71	31.82	7.21	5.625	12.7	12.84	8.03	7.21	3.62	4.50	2160.59	3.61	0.007321	4.45
674	95	7.4	5.76	31.83	7.61	5.938	12.9	13.55	8.47	7.61	3.68	4.54	2236.96	3.67	0.007321	4.48
680	100	7.1	5.81	31.85	8.01	6.250	13.2	14.26	8.92	8.01	3.76	4.56	2311.72	3.76	0.007321	4.50
686	105	6.9	5.86	31.87	8.40	6.406	13.4	14.81	9.21	8.40	3.82	4.59	2359.49	3.81	0.007321	4.53
693	110	6.7	5.92	31.89	8.80	6.875	13.6	15.67	9.81	8.80	3.88	4.63	2457.43	3.87	0.007321	4.57
702	115	6.4	6.00	31.91	9.19	7.188	13.9	16.38	10.25	9.19	3.96	4.68	2528.45	3.96	0.007321	4.61

Note sample number: 3–5. Test date: April 28, 2002. Calibration coefficient of force ring: 2.55 kg/0.01 m. Initial height: 11.7 cm. Initial reading of drain pipe: 20.3 cm^3. Loading rate: 1.0 mm/min

Table 5.7 Triaxial test data with equal principal stress ratio $k = 0.6$

Axial deformation (μm)	Measure the force ring (mm)	Drain tube reading (mm)	Axial strain (%)	Correct the area (mm²)	Main stress difference (kPa)	surrounding pressure (kPa)	water discharge (m)	Major principal stress (kPa)	K	P (kPa)	Q (kPa)	Volume strain (%)	Shear strain (%)	K	Plastic volume deformation (%)	1/G (mm/N)	Plastic shear (%)
255	20	16.5	1.98	30.60	1.67	2.600	9.7	4.27	0.61	3.16	1.67	2.51	1.14	1184.34	2.50	0.01379	1.12
262	22.5	16	2.03	30.62	1.87	2.925	10.2	4.80	0.61	3.55	1.87	2.64	1.15	1277.54	2.63	0.01379	1.13
260	20	16.2	2.02	30.62	1.67	2.600	10.0	4.27	0.61	3.16	1.67	2.58	1.15	1184.28	2.58	0.01379	1.13
259	17.5	16.4	2.01	30.61	1.46	2.275	9.8	3.73	0.61	2.76	1.46	2.53	1.16	1086.75	2.53	0.01379	1.14
256	15	16.6	1.98	30.61	1.25	1.950	9.6	3.20	0.61	2.37	1.25	2.48	1.16	984.12	2.48	0.01379	1.14
254	12.5	16.8	1.97	30.60	1.04	1.630	9.4	2.67	0.61	1.98	1.04	2.43	1.16	876.59	2.43	0.01379	1.14
250	10	17.1	1.94	30.59	0.83	1.300	9.1	2.13	0.61	1.58	0.83	2.35	1.15	758.09	2.35	0.01379	1.14
247	7.5	17.5	1.91	30.59	0.63	0.975	8.7	1.60	0.61	1.18	0.63	2.25	1.17	629.96	2.25	0.01379	1.16
242	5	18	1.88	30.57	0.42	0.650	8.2	1.07	0.61	0.79	0.42	2.12	1.17	485.27	2.12	0.01379	1.16
245	7.5	17.6	1.90	30.58	0.63	0.975	8.6	1.60	0.61	1.18	0.63	2.22	1.16	629.97	2.22	0.01379	1.15
248	10	17.2	1.92	30.59	0.83	1.300	9.0	2.13	0.61	1.58	0.83	2.33	1.15	758.11	2.32	0.01379	1.14
251	12.5	16.9	1.95	30.60	1.04	1.630	9.3	2.67	0.61	1.98	1.04	2.40	1.14	876.61	2.40	0.01379	1.13
255	15	16.5	1.98	30.60	1.25	1.950	9.7	3.20	0.61	2.37	1.25	2.51	1.14	984.13	2.50	0.01379	1.12
258	17.5	16.3	2.00	30.61	1.46	2.275	9.9	3.73	0.61	2.76	1.46	2.56	2.56	1086.76	2.56	0.01379	1.13
261	20	16	2.02	30.62	1.67	2.600	10.2	4.27	0.61	3.16	1.67	2.64	2.64	1184.27	2.63	0.01379	1.12
265	22.5	15.8	2.05	30.63	1.87	2.925	10.4	4.80	0.61	3.55	1.87	2.69	2.69	1277.51	2.68	0.01379	1.13
271	25	15.5	2.10	30.64	2.08	3.250	10.7	5.33	0.61	3.94	2.08	2.77	1.18	1367.08	2.56	0.01379	1.15
277	27.5	15	2.15	30.66	2.29	3.575	11.2	5.86	0.61	4.34	2.29	2.89	1.18	1453.50	2.63	0.01379	1.15
283	30	14.7	2.19	30.67	2.49	3.900	11.5	6.39	0.61	4.73	2.49	2.97	1.20	1537.15	2.68	0.01379	1.17
289	32.5	14.4	2.24	30.69	2.70	4.225	11.8	6.93	0.61	5.13	2.70	3.05	1.22	1618.34	2.76	0.01379	1.19
295	35	14.1	2.29	30.70	2.91	4.550	12.1	7.46	0.61	5.52	2.91	3.13	1.24	1697.32	2.89	0.00643	1.23
302	37.5	13.7	2.34	30.72	3.11	4.875	12.5	7.99	0.61	5.91	3.11	3.23	1.26	1774.28	2.97	0.00643	1.24

(continued)

Table 5.7 (continued)

Axial deformation (μm)	Measure the force ring (mm)	Drain tube reading (mm)	Axial strain (%)	Correct the area (mm^2)	Main stress difference (kPa)	surrounding pressure (kPa)	water discharge (m)	Major principal stress (kPa)	*K*	*P* (kPa)	*Q* (kPa)	Volume strain (%)	Shear strain (%)	*K*	Plastic volume deformation (%)	1/G (mm/N)	Plastic shear (%)
308	40	13.3	2.39	30.73	3.32	5.200	12.9	8.52	0.61	6.31	3.32	3.33	1.28	1849.45	3.05	0.00643	1.25
314	42.5	13	2.43	30.75	3.52	5.525	13.2	9.05	0.61	6.70	3.52	3.41	1.30	1922.94	3.41	0.00643	1.27
319	45	12.8	2.47	30.76	3.73	5.850	13.4	9.58	0.61	7.09	3.73	3.46	1.32	1994.92	3.46	0.00643	1.29
324	47.5	12.5	2.51	30.77	3.94	6.175	13.7	10.11	0.61	7.49	3.94	3.54	1.33	2065.48	3.54	0.00643	1.31
324	45	12.5	2.51	30.77	3.73	5.850	13.7	9.58	0.61	7.09	3.73	3.54	1.33	1994.83	3.54	0.00643	1.31
323	42.5	12.6	2.50	30.77	3.52	5.525	13.6	9.05	0.61	6.70	3.52	3.51	1.33	1922.79	3.51	0.00643	1.31
322	40	12.7	2.50	30.77	3.32	5.200	13.5	8.52	0.61	6.31	3.32	3.49	1.33	1849.22	3.49	0.00643	1.31
321	37.5	12.8	2.49	30.77	3.11	4.875	13.4	7.98	0.61	5.91	3.11	3.46	1.33	1773.98	3.46	0.00643	1.31
318	35	13	2.47	30.76	2.90	4.550	13.2	7.45	0.61	5.52	2.90	3.41	1.33	1696.97	3.41	0.00643	1.31
315	30	13.2	2.44	30.75	2.49	3.900	13	6.39	0.61	4.73	2.49	3.36	1.32	1536.71	3.36	0.00643	1.31
312	25	13.5	2.42	30.74	2.07	3.250	12.7	5.32	0.61	3.94	2.07	3.28	1.32	1366.58	3.28	0.00643	1.31
307	20	13.9	2.38	30.73	1.66	2.600	12.3	4.26	0.61	3.15	1.66	3.18	1.32	1183.79	3.18	0.00643	1.31
300	15	14.3	2.33	30.71	1.25	1.950	11.9	3.20	0.61	2.37	1.25	3.08	1.30	983.73	3.07	0.00643	1.29
294	10	14.9	2.28	30.70	0.83	1.300	11.3	2.13	0.61	1.58	0.83	2.92	1.31	757.79	2.92	0.00643	1.30
298	15	14.2	2.31	30.71	1.25	1.950	12	3.20	0.61	2.37	1.25	3.10	1.28	983.75	3.10	0.00643	1.27
303	20	13.9	2.35	30.72	1.66	2.600	12.3	4.26	0.61	3.15	1.66	3.18	1.29	1183.83	3.18	0.00643	1.28
308	25	13.4	2.39	30.73	2.07	3.250	12.8	5.32	0.61	3.94	2.07	3.31	1.28	1366.63	3.31	0.00643	1.27
313	30	13.0	2.43	30.75	2.49	3.900	13.2	6.39	0.61	4.73	2.49	3.41	1.29	1536.74	3.41	0.00643	1.27
318	35	12.7	2.47	30.76	2.90	4.550	13.5	7.45	0.61	5.52	2.90	3.49	1.30	1696.97	3.49	0.00643	1.28
323	40	12.5	2.50	30.77	3.31	5.200	13.7	8.51	0.61	6.30	3.31	3.54	1.32	1849.20	3.54	0.00643	1.30
329	45	12.3	2.55	2.55	3.73	3.73	13.9	9.58	0.61	7.09	3.73	3.59	1.35	1994.74	3.59	0.00643	1.33
536	50	12	4.16	4.16	4.07	4.07	14.2	10.57	0.61	7.86	4.07	3.67	2.93	2130.76	3.67	0.00643	2.91

(continued)

Table 5.7 (continued)

Axial deformation (μm)	Measure the force ring (mm)	Drain tube reading (mm)	Axial strain (%)	Correct the area (mm^2)	Main stress difference (kPa)	surrounding pressure (kPa)	water discharge (m)	Major principal stress (kPa)	*K*	*P* (kPa)	*Q* (kPa)	Volume strain (%)	Shear strain (%)	*K*	Plastic volume deformation (%)	1/G (mm/N)	Plastic shear (%)
544	55	11.5	4.22	4.22	4.48	4.48	14.7	11.63	0.61	8.64	4.48	3.80	2.95	2265.41	3.80	0.00643	2.92
555	60	11	4.30	4.30	4.88	4.88	15.2	12.68	0.62	9.43	4.88	3.93	2.99	2395.68	3.92	0.00643	2.96
563	65	10.6	4.36	4.36	5.28	5.28	15.6	13.73	0.62	10.21	5.28	4.03	3.02	2522.16	4.03	0.00643	2.99

Note Sample number: 3–3. Test date: April 26, 2002. Force ring correction factor: 2.55 kg/0.01 m. Initial height: 12.9 cm. Drain initial reading: 26.2 cm^3. Loading rate: 1.0 mm/min

Table 5.8 Hydrostatic test data

Drain tube readings (mm)	surrounding pressure (kPa)	Water discharge (mL)	Body strain (%)	lg *p* (kPa)	K_v
47.6	0.00	0	0.00	/	0.38
45.6	0.10	2	054	-100	
439	020	3.7	1.01	-0.70	
42.7	030	49	135	-052	
413	040	63	174	-040	
39.9	0.50	7.7	213	-030	
39	0.60	8.6	239	-0.22	
382	0.70	94	261	-0.15	
37.5	0.80	10.1	281	-0.10	
37.1	0.90	105	2.93	-0.05	
36.8	100	10.8	3.02	0.00	
369	090	10.7	2.99	-0.05	
37	0.80	10.6	296	-0.10	
37.1	0.70	105	2.93	-0.15	
372	060	10.4	2.90	-022	
37.3	0.50	103	2.87	-0.30	
37.4	0.40	102	284	-0.40	
376	030	10	2.79	-0.52	
37.8	020	9.8	2.73	-0.70	
381	0.10	9.5	2.64	-100	
37.8	0.20	9.8	2.73	-0.70	
37.6	030	10	2.79	-052	
372	050	10.4	2.90	-0.30	
36.8	0.70	10.8	3.02	-0.15	
36.5	0.90	111	3.10	-0.05	
363	100	113	3.16	0.00	
359	120	117	3.27	0.08	
35.6	140	12	3.36	0.15	
352	1.60	124	3.48	020	
349	1.80	12.7	356	0.26	
34.6	2.00	13	3.65	0_30	
34.7	1.80	129	3.62	026	
34.85	1.60	12.75	3.58	020	
35	1.40	12.6	3.54	0.15	
35.1	120	125	351	0.08	

(continued)

Table 5.8 (continued)

Drain tube readings (mm)	surrounding pressure (kPa)	Water discharge (mL)	Body strain (%)	lg p (kPa)	K_v
35.3	1.00	12.3	3.45	0.00	0.61
35.55	0.80	12.05	3.38	− 0.10	
35.8	0.60	11.8	3.30	− 0.22	
36.2	0.40	11.4	3.19	− 0.40	
36.7	0.20	10.9	3.04	− 0.70	
36.25	0.40	11.35	3.17	− 0.40	
35.95	0.60	11.65	3.26	− 0.22	
35.65	0.80	11.95	3.35	− 0.10	
35.4	1.00	12.2	3.42	0.00	
35.2	1.20	12.4	3.48	0.08	
35	1.40	12.6	3.54	0.15	
34.8	1.60	12.8	3.59	0.20	
34.5	1.80	13.1	3.68	0.26	
34.2	2.00	13.4	3.77	0.30	
33.9	2.20	13.7	3.86	0.34	
33.7	2.40	13.9	3.91	0.38	0.55
33.5	2.60	14.1	3.97	0.41	
33.3	2.80	14.3	4.03	0.45	
33.1	3.00	14.5	4.09	0.48	
33.2	2.70	14.4	4.06	0.43	
33.3	2.40	14.3	4.03	0.38	
33.4	2.10	14.2	4.00	0.32	
33.6	1.80	14	3.94	0.26	
33.8	1.50	13.8	3.89	0.18	
34	1.20	13.6	3.83	0.08	
34.3	0.90	13.3	3.74	− 0.05	
34.7	0.60	12.9	3.62	− 0.22	
35.3	0.30	12.3	3.45	− 0.52	
35.9	0.10	11.7	3.27	− 1.00	
35.2	0.30	12.4	3.48	− 0.52	
34.7	0.60	12.9	3.62	− 0.22	
34.2	0.90	13.4	3.77	− 0.05	
33.9	1.20	13.7	3.86	0.08	
33.5	1.50	14.1	3.97	0.18	
33.3	1.80	14.3	4.03	0.26	

(continued)

Table 5.8 (continued)

Drain tube readings (mm)	surrounding pressure (kPa)	Water discharge (mL)	Body strain (%)	lg *p* (kPa)	K_v
32.95	2.20	14.65	4.13	0.34	
32.7	2.50	14.9	4.21	0.40	
32.5	2.80	15.1	4.27	0.45	
32.1	3.20	15.5	4.38	0.51	
31.8	3.60	15.8	4.47	0.56	
31.4	4.00	16.2	4.59	0.60	
31.5	3.60	16.1	4.56	0.56	0.7
31.6	3.20	16	4.53	0.51	
31.7	2.80	15.9	4.50	0.45	
32	2.40	15.6	4.41	0.38	
32.2	2.00	15.4	4.36	0.30	
32.5	1.60	15.1	4.27	0.20	
32.85	1.20	14.75	4.16	0.08	
33.3	0.80	14.3	4.03	−0.10	
33.6	0.60	14	3.94	−0.22	
33.95	0.40	13.65	3.84	−0.40	
34.5	0.20	13.1	3.68	−0.70	
34.1	0.40	13.5	3.80	−0.40	
33.8	0.60	13.8	3.89	−0.22	
33.5	0.80	14.1	3.97	−0.10	
33.2	1.00	14.4	4.06	0.00	
33	1.20	14.6	4.12	0.08	
32.6	1.60	15	4.24	0.20	
32.3	2.00	15.3	4.33	0.30	
32	2.40	15.6	4.41	0.38	
31.8	2.80	15.8	4.47	0.45	
31.45	3.30	16.15	4.58	0.52	
31.15	3.80	16.45	4.67	0.58	0.69
30.8	4.40	16.8	4.77	0.64	
30.2	5.00	17.4	4.95	0.70	
30.35	4.50	17.25	4.90	0.65	
30.5	4.00	17.1	4.86	0.60	
30.7	3.50	16.9	4.80	0.54	
30.85	3.00	16.75	4.76	0.48	
31.1	2.50	16.5	4.68	0.40	

(continued)

Table 5.8 (continued)

Drain tube readings (mm)	surrounding pressure (kPa)	Water discharge (mL)	Body strain (%)	lg p (kPa)	K_v
31.35	2.00	16.25	4.61	0.30	
31.7	1.50	15.9	4.50	0.18	
32.2	1.00	15.4	4.36	0.00	
32.9	0.50	14.7	4.15	−0.30	
33.4	0.30	14.2	4.00	−0.52	
34.2	0.10	13.4	3.77	−1.00	
33.6	0.30	14	3.94	−0.52	
33.15	0.50	14.45	4.08	−0.30	
32.4	1.00	15.2	4.30	0.00	
31.9	1.50	15.7	4.44	0.18	
31.5	2.00	16.1	4.56	0.30	
31.1	2.50	16.5	4.68	0.40	
30.9	3.00	16.7	4.74	0.48	
30.6	3.50	17	4.83	0.54	
30.3	4.00	17.3	4.92	0.60	
30	4.50	17.6	5.01	0.65	
29.8	5.00	17.8	5.07	0.70	
29.5	5.50	18.1	5.16	0.74	
29.2	6.00	18.4	5.25	0.78	
29	6.50	18.6	5.31	0.81	

Note Sample number: 0–1. Test date: July 20, 2002. Force ring correction factor: 2.523 kg/0.01 m. Initial height: 12.3 cm. Drain initial reading: 47.6 cm^3. load rate: 1.0 mm/min

Assuming only elastic deformation occurs when unloading, the lg$p$-ε_v curve is obtained based on the experimental results shown in Table 5.8, as shown in Fig. 5.2. Assuming that the bounce line is a straight line, then

$$\varepsilon_v = \varepsilon_{vl} + k_v \lg p \tag{5.8}$$

where, ε_{vl} is the volume deformation when P = 100 kPa; k_v is the slope of the rebound line, $k_v = tan\alpha$. The Formula (5.8) is divided on the rebound line, $d\varepsilon_v = d\varepsilon_v^e$,$d\varepsilon_v^e = k_v \times 0.434dp/p$. The elastic volume compression modulus K is

$$K = \frac{dp}{d\varepsilon_v^e} = \frac{p}{0.434k_v} \tag{5.9}$$

From the lg$p$-ε_v curve, the slope of spring back line at p = 100 kPa, 200 kPa, 300 kPa, 400 kPa and 500 kPa is k_v = 0.38, 0.61, 0.55, 0.7 and 0.67, respectively.

Fig. 5.2 lgp-εv curve

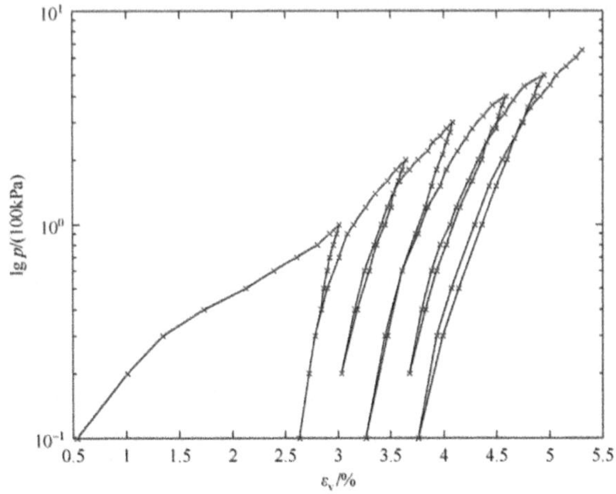

Then the K value is 606.4p, 755.5p, 1256.8p, 1316.7p and 1669.7p, respectively. If the K and p is plotted on a double logarithmic curve, which is approximately a straight line, as shown in Fig. 5.3, then K can be written as

$$K = K_1 p_a \left(\frac{p}{p_a} \right)^n \tag{5.10}$$

where, $Pa = 100$ kPa. when $p = 100$ kPa, $k_1 = 565.234$, n is the slope of the straight line, $n = 0.6437$.

$$K = 565.234 p^{0.6437} \tag{5.11}$$

Fig. 5.3 The curves of K-p

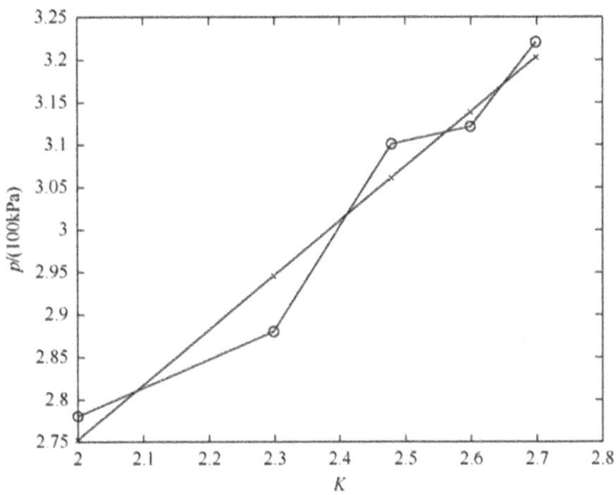

5.4 Stress Strain Curve of Triaxial Test with Equal Principal Stress Ratio

Figures 5.4, 5.5, 5.6, 5.7 and 5.8 show the corresponding $\varepsilon_v^p - q$ and $\overline{\varepsilon}^p - q$ curves of specimens with equal principal stress ratios $k = 0.3, 0.42, 0.5, 0.6$ and 0.7

5.5 Three Dimensional Surface of Stress Strain

By drawing the shear strain ε^-, average normal stress p and generalized shear stress q under three stress paths together, the stress–strain curve in the whole stress field (p, q) can be obtained, as shown in Figs. 5.9, 5.10 and 5.11. The three-dimensional surface graph is more vivid and intuitive than the two-dimensional graph, and can

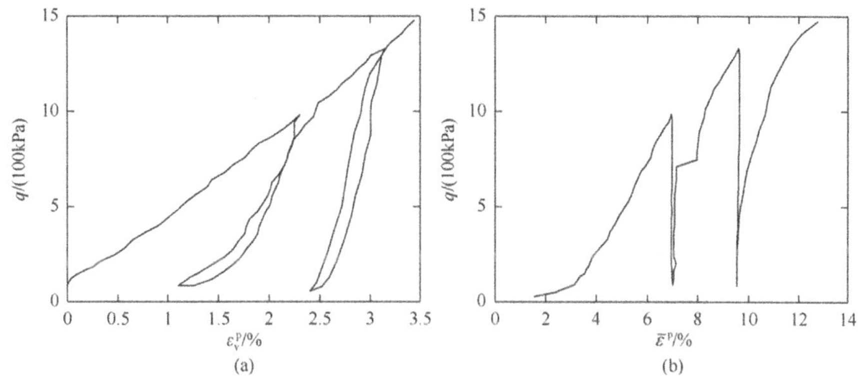

Fig. 5.4 Stress strain curve at k = 0.3

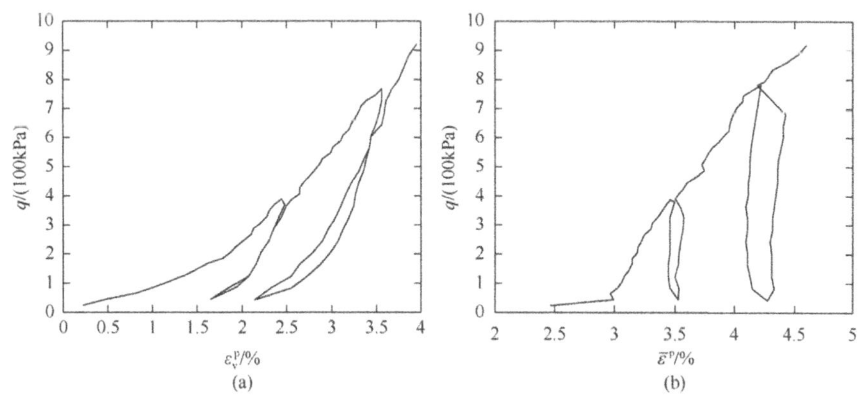

Fig. 5.5 Stress strain curve at k = 0.42

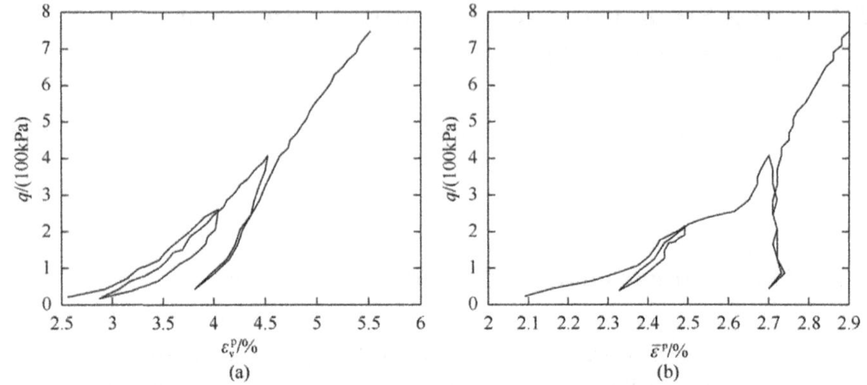

Fig. 5.6 Stress strain curve at k = 0.5

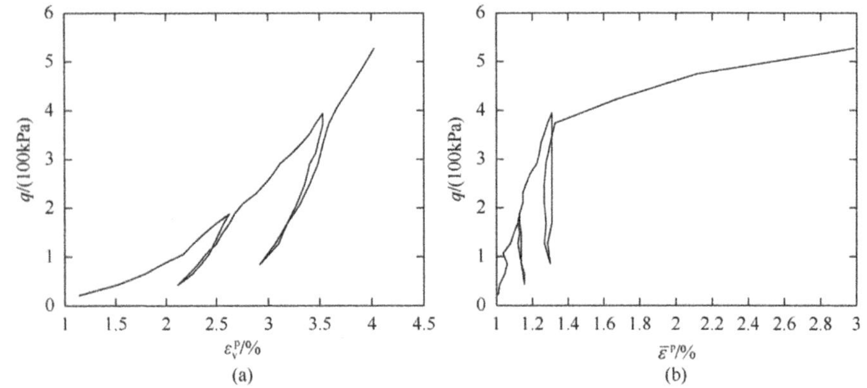

Fig. 5.7 Stress strain curve at k = 0.6

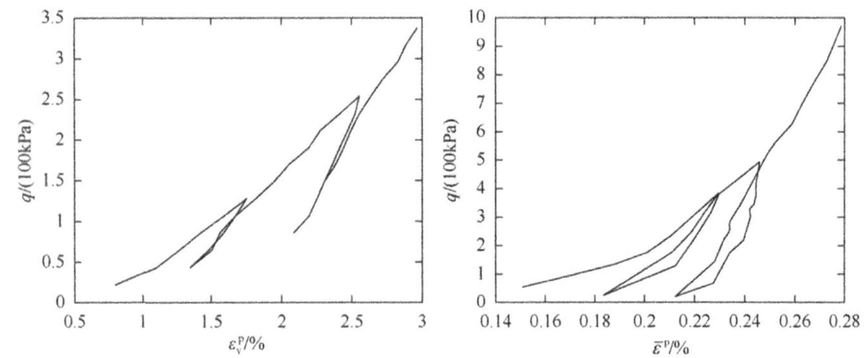

Fig. 5.8 Stress strain curve at k = 0.7

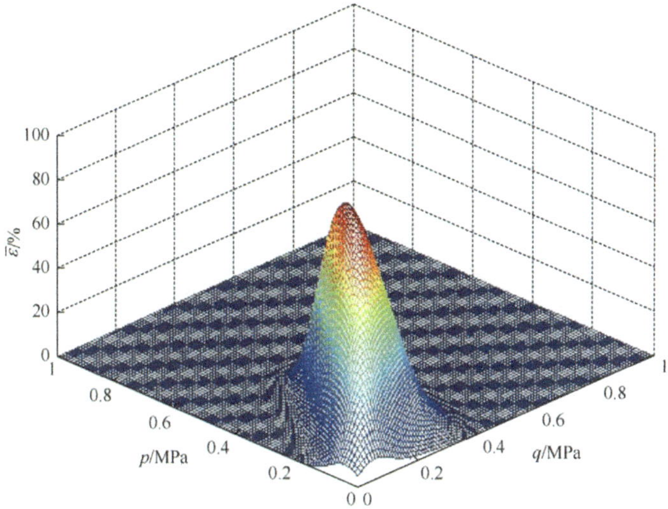

Fig. 5.9 Total stress–strain curve of equal principal stress ratio

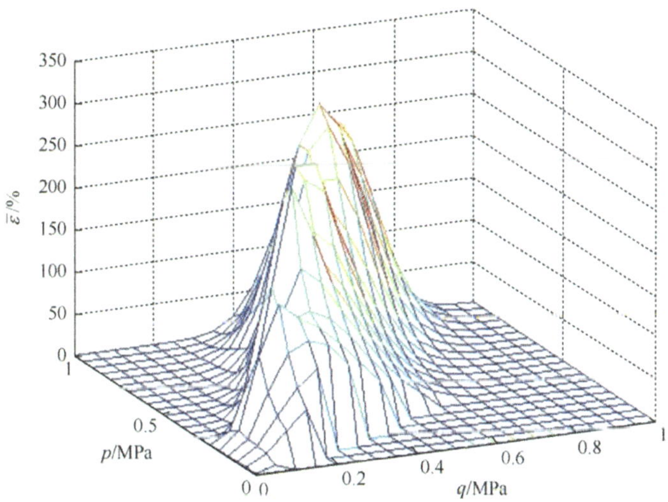

Fig. 5.10 General path full stress–strain curve of conventional

fully describe the constitutive relationship of soil. From the comparison of the stress–strain curves under the equal principal stress ratio path, the conventional path and the equal p path, it can be seen that the peak value of the surface under the conventional path is the largest, and the equal p path is the smallest. The peak value of the surface under the path is the smallest, and the peak value of the surface under the equal principal stress ratio path is between the two, which indicates that the shear strain of the soil sample is the largest at the maximum failure stress, reflecting the compressive

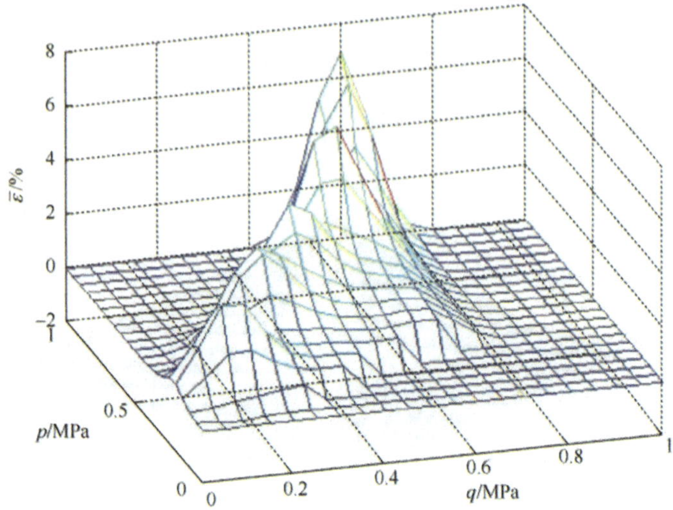

Fig. 5.11 Total stress–strain curve with equal p-path

strength of the soil. At the same time, the surface under the conventional path is the steepest, that is, the strain changes fastest with the stress, and the equal p. The surface under the path is gentle, and the surface under the equal principal stress ratio path is between them. From the above comparison, it is obvious that the influence of different stress paths on the stress–strain relationship of soil is significant, and after the treatment of this book, this difference can not only be observed qualitatively, but also be given quantitatively, which provides a reliable basis for the study of the influence of stress paths on the constitutive relationship in the future.

5.6 Yield Trajectory

It is considered that the yield trajectory can be drawn directly from the test data. The volume yield line with plastic volume strain as hardening parameter and the shear yield line with as hardening parameter are drawn from the constitutive model shown in Eqs. (3.1) and (3.2). The specific method is to find out the corresponding points of and with equal values from the and values corresponding to any point M (p, q) on the test curve, and draw the corresponding yield trajectory contour of $H = \varepsilon_v^p$ and $H = \bar{\varepsilon}^p$ work hardening law from the p and q values of these points, as shown in Figs. 5.12, 5.13 and 5.14.

It can be seen from Figs. 5.12, 5.13 and 5.14 that on the one hand, in the range of yield line, the conventional path is the largest, the equal p path is the smallest, and the equal principal stress ratio path is in the middle [2–5]. The results show that the anti-failure ability of the soil under the conventional path is the strongest. The

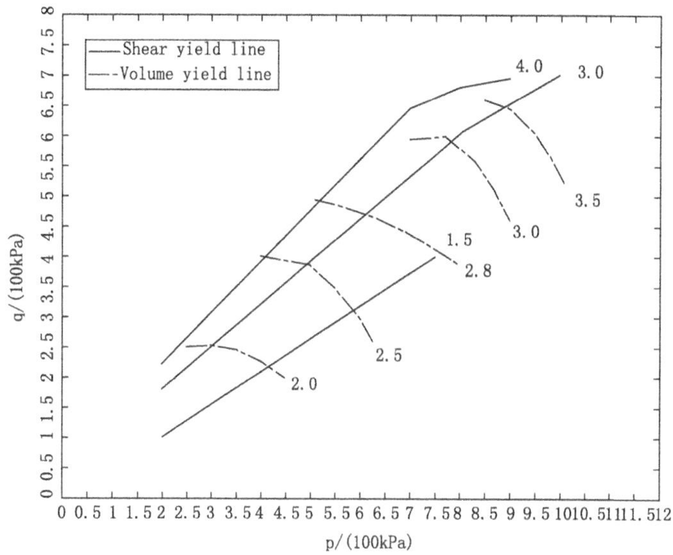

Fig. 5.12 Yield path of equal principal stress ratio

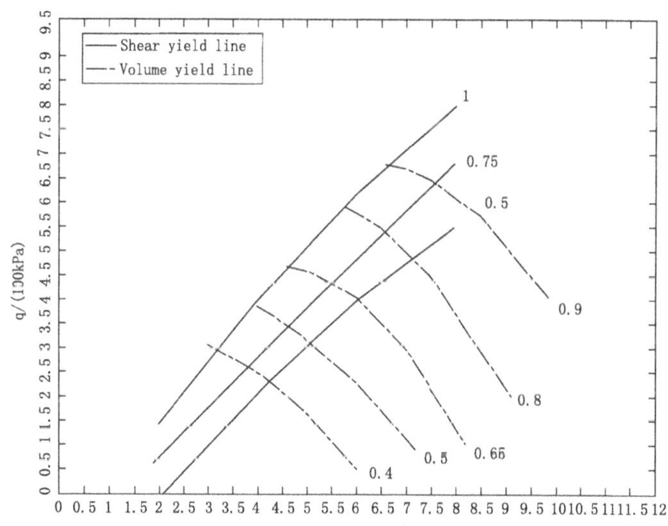

Fig. 5.13 Yield trajectory of conventional path

anti-failure ability of the soil under the equal p path is the smallest, which reflects the compaction of the excavated soil again [6, 7]. On the other hand, no matter the volume yield line or the shear yield line, the slope of the conventional path is the largest, and the slope of the equal p path is the smallest, and the slope of the equal principal stress ratio path is between them. This result is consistent with the analysis

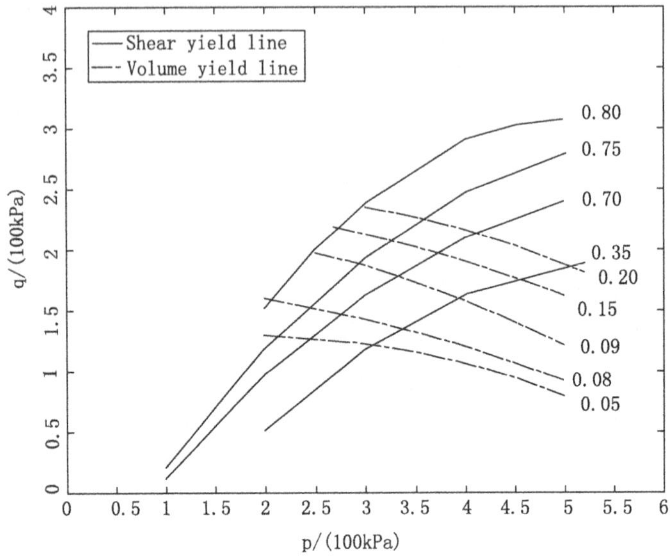

Fig. 5.14 Yield trajectory of equal p path

result of full stress–strain curve of soil in Sect. 5.5. It is further proved that the effect of stress path on soil constitutive relation can not be ignored.

5.7 Establishment of Numerical Model

5.7.1 Constitutive Model

The sand is similar to expansive soil. The stress–strain relationship of sand is also expressed by the following two functionals.

$$\varepsilon_v = f_1(p, q, p(q)) \tag{5.12}$$

$$\overline{\varepsilon} = f_2(p, q, p(q)) \tag{5.13}$$

where, εv and $\overline{\varepsilon}$ are volume strain and octahedral shear strain, respectively. p and q are mean normal stress and general shear stress, respectively. $p(q)$ is stress path function. f_1 and f_2 are two arbitrary nonlinear functions.

The stress–strain relationship reflects the coupling relationship between volume strain and shear strain and the cross effect of p and q. It not only considers the dilatancy and shrinkage of soil, but also reflects the effect of stress path on the constitutive relationship. Because the model parameters in this book are obtained by experimental simulation according to the actual stress path of soil [8–12]. Therefore, the effect of stress path on the constitutive relationship has been considered. So the above formula can be simplified as

$$\varepsilon_v = f_1(p, q) \tag{5.14}$$

$$\overline{\varepsilon} = f_2(p, q) \tag{5.15}$$

The incremental form of the above two formulas is shown in the following formula.

$$d\varepsilon_v = \frac{\partial f_1(p, q)}{\partial p}dp + \frac{\partial f_1(p, q)}{\partial q}dq \tag{5.16}$$

$$d\overline{\varepsilon} = \frac{\partial f_2(p, q)}{\partial p}dp + \frac{\partial f_2(p, q)}{\partial q}dq \tag{5.17}$$

The coefficients $\partial f_1/\partial p$, $\partial f_1/\partial q$,$\partial f_2/\partial p$ and $\partial f_2/\partial q$ of the above four equations are functions of p and q, which can be determined by triaxial test. When the same stress state (p, q) is reached through different stress paths in the test, the values of the above four coefficients can be different, so as to reflect the influence of stress path. The four coefficients can be obtained directly from the curve obtained from the experiment.

5.7.2 Neural Network Learning and Prediction

RBF neural network is used for training [13–16]. The average normal stress p and average shear stress q are input vectors. The volume strain εv and shear strain $\overline{\varepsilon\varepsilon}$ arc output vectors [17, 18]. The number of neurons in hidden layer is calculated by empirical formula [19]. For each group of test samples, the test data p, q, ε_v and $\overline{\varepsilon\varepsilon}$ with equal principal stress ratios $k = 0.42$, 0.5 and 0.7 are taken as the learning objects. Then the trained neural network is used to predict the actual test data under the path $k = 0.6$ to test the accuracy of the network model.

The predicted results for the path with equal principal stress ratio $k = 0.6$ are shown in Fig. 5.15. It can be seen from the Fig. 5.15 that the predicted results of the neural network are very close to the actual experimental data, which verifies once again that the neural network has a powerful function in establishing the numerical model of geotechnical constitutive relationship, and can truly reflect the stress–strain state of the sample.

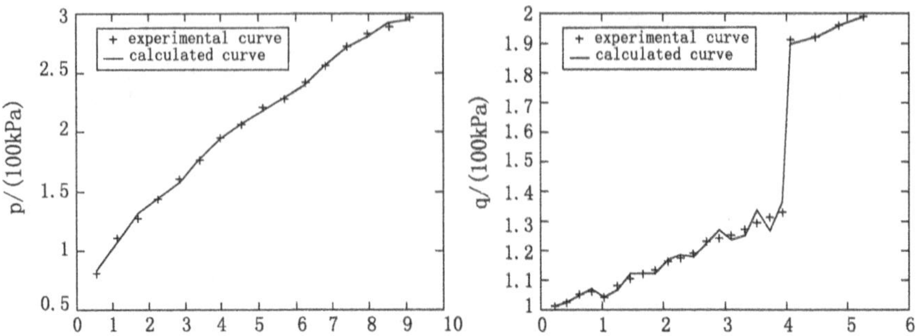

Fig. 5.15 Comparison of network predicted value and experimental value

5.8 Validation of Numerical Model

The sand samples are verified by the same method as the expansive soil in Chap. 4. The calculated data at $k = 0.5$ are compared with the experimental data, as shown in Fig. 5.16. The solid line in the Fig. 5.16 is the test curve, and the dotted line is the calculation curve. It can be seen from the Fig. 5.16 that the calculated results are consistent with the test data, which indicates that the numerical model of sand constitutive relationship established in this book is correct and can accurately reflect the stress–strain relationship of sand under different stress paths.

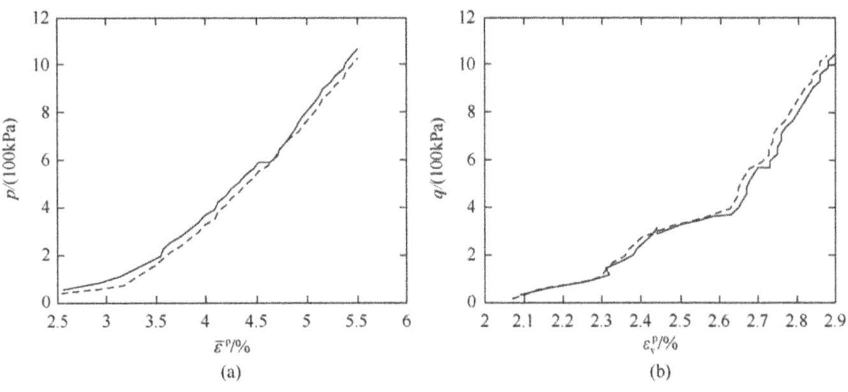

Fig. 5.16 Comparison of calculated and experimental values

References

1. Dong Q, Yao H, Lu Z et al (2014) Study of microfabric and constitutive relation of granular material. Chin J Rock Mechan Eng 33(2):4334–4342
2. Kong L, Zheng Y, Wang Y (2000) A three-yield-surface model for soilmasses based on the generalized plastic mechanics. Rock Soil Mech 21(2):108–112
3. Li Q (2001) Application of ann and probability method to information technology for construction of deep excavation. Chin Civil Eng J 34(2):96–100
4. Lin F (2007) Study on neural network model for constitutive relationship of coarse grained soil. Tianjin University, Tianjin
5. Lu Z, Gong X (1997) The proof that the yield curve of the yield condition of the stable material in the two plane is captured in the inner and outer envelope. Acta Geotech Eng Sinica 19(5):1–5
6. She Y, Liu H, Gao Y (2003) Neural networks method for computation of autocovariance distance. Rock Soil Mech 24(5):719–722
7. Dean S, Yangping Y, Zongze Y (1999) Generalization of elasto-plastic model with two yield surfaces based on SMP criterion. Chin J Geotech Eng 21(5):631–634
8. Tan Y, Wang C (2001) A fast approaching model for rock constitutive equation by radial basis function neural network. Chin J Geotech Eng 23(1):14–19
9. Guo Z (2001) A radial basis function artificial neural network model for water quality evaluation. J Zhejiang Uni 27(3):335–338
10. He X, Li S, Liu Y, et al (2003) Artificial neural network method for stability prediction of geotechnical slopes. Rock Soil Mech 24:73–76
11. Wang C, Su J, Lin F (2013) A study of the BP neural network constitutive model for granular soils
12. Akiyoshi T, Matsumoto H, Fuchida K et al (1995) Cyclic mobility behaviour of sand by the three-dimensional strain space multi-mechanism model. Int J Rock Mech Mining Sci Geomech Abstr 32(4):155–174
13. Ellis GW, Yao C, Zhao R et al (1995) Stress-strain modeling of sands using artificial neural networks. J Geotech Eng 121(5):429–435
14. Fahlman SE (1988) An empirical study of learning speed in back-propagation networks. Tech Rep CMU-CS-88-162 (9):112–119
15. Lau HCW, Ning A, Pun KF et al (2000) Neural network for the dimensional control of modeled parts based on a reverse process model. J Mater Process Technol 117:89–96
16. Qian HC, Xia BC, Li SZ et al (2002) Fuzzy neural network modeling of material properties. J Mater Process Technol 122:196–200
17. Shahin MA, Jaksa MB, Maier HR (2001) Artificial neural network applications in geotechnical engineering. Austr Geotech: 49–62
18. Szecsi T (1999) Cutting force modeling using artificial neural networks. J Mater Process Technol 92:344–349
19. Toshio F, Takanori G (1992) Theory and application of neural networks for industrial control system. IEEE Trans Ind Electron 12:1001–1013

Chapter 6
Triaxial Test and Numerical Modeling of Clay

6.1 Sample Preparation

The 10 kg representative soil sample of foundation pit was selected, dried, crushed and screened by 2 mm. The moisture content of air drying is 5.92%, and the required water content is 33.4%. The required amount of water on the soil material sprayed and mixed well. Then, it is put into a plastic bag after standing slightly. Finally, it is put into a closed container for 24 h to make the water content uniform. The soil material and retest its moisture content are taken out. The difference between the measured moisture content and the required moisture content should be less than ±1%. Otherwise, the moisture content adjusted, until it meets the requirements. According to the required dry density of 1.44 g/cm^3, the required wet soil mass is weighed. According to the sample height, it is compacted in 6 layers. The quality of each layer is required to be equal. Each layer shall be compacted to the required height. Roughen the surface. Then, the second layer of soil is added. This continues until the last layer is hit. Both ends of the sample are placed in the sample barrel, taken out and weighed. The density difference of the sample should be less than 0.02 g/cm^3. The qualified standard of sample is 183.2–185.8 g after preparation, otherwise it is a waste sample.

Through the above sample preparation process, it can be seen that the quality control standards of remolded soil samples are moisture content, and dry density. Both of them must meet the requirements at the same time before they are qualified. In fact, in the process of sample preparation in strict accordance with the above standards, after repeated, the waste sample rate is also high, the test samples used in this book are qualified samples. According to the limit moisture content, specific gravity of soil particles and other tests, the physical property indexes and shape parameters of the sample are shown in Table 6.1.

Q. Ren and J. Zhou, *Numerical Modeling of Soil Constitutive Relationship*,
https://doi.org/10.1007/978-981-16-3231-0_6

Table 6.1 Physical indicators and shape parameters of the test

17 mm liquid limit (%)	10 mm liquid limit (%)	Plastic limit (%)	Plasticity index	Moisture content (%)	Dry density (g/cm^3)	Proportion of soil particles	Sample height (cm)	Sample diameter (cm)
48.19	42.67	29.52	18.67	33.4	1.44	2.73	8	3.91

6.2 Sample Saturation

6.2.1 Pumping Saturation

The sample is put into the saturator, and then put into the anhydrous vacuum cylinder. The vacuum cylinder is connected with the extractor for air extraction. When the true void is close to 1 atmospheric pressure, continue to pump for 2 h. Then, water is slowly injected into the vacuum cylinder to keep the vacuum stable. When the saturator is completely flooded, stop pumping and release the vacuum in the vacuum cylinder. The sample is allowed to stand under water for 24 h, and then the sample is taken out and weighed.

6.2.2 Back Pressure Saturation

(1) After the sample pressure chamber is installed on the base, the valve chamber is closed and the pore pressure is recorded. Firstly, the 20 kPa surrounding pressure is applied to the sample. The pore pressure valve is opened. After the pore pressure is stable, the reading is recorded. Then, the pore pressure valve is closed.

(2) The back pressure is applied step by step. The surrounding pressure is applied step by step at the same time, so as to minimize the disturbance to the sample. In the process of applying back pressure, always keep the surrounding pressure 20 kPa higher than the back pressure. The increment of each stage of back pressure and surrounding pressure is 50 kPa.

(3) During operation, firstly the ambient pressure is adjusted to 50 kPa, then the back pressure system is adjusted to 30 kPa. At the same time, the ambient pressure valve and back pressure valve is opened. Then the pore pressure valve is slowly opened. After the pore pressure is stable, the readings of pore pressure gauge and volume transformer tube measured and recorded. Then the next level of ambient pressure and back pressure is applied.

(4) The pore pressure increment Δu under the ambient pressure calculated at the same level. compared it with the ambient pressure increment $3\Delta\sigma$. If $3\Delta u/\Delta\sigma < 1$, it means that the sample is not saturated. At this time, the pore pressure valve is closed. Back pressure valve and ambient pressure valve are continuously

applied to the next level of ambient pressure and back pressure, according to the above provisions.

(5) When the sample reaches $3\Delta u/\Delta\sigma = 1$ at a certain pressure level, keep the back pressure unchanged and increase the surrounding pressure. If the increased pore pressure in the sample is equal to the increment of the surrounding pressure, it indicates that the sample is completely saturated. Otherwise, the above steps are repeated until the sample is saturated.

6.3 Specimen Installation and Consolidation

The procedure of specimen installation and consolidation is as follows.

(1) The pore pressure valve and the measuring pipe valve are opened to make the pressure chamber base filled with water and exhaust, and close the valve. The boiled permeable plate is slid onto the base of the pressure chamber. Then put the wet filter paper and the sample, and put a wet filter paper and the permeable plate on the top of the sample.

(2) The rubber membrane is put into the membrane bearing cylinder. Its two ends are screwed out of the cylinder. Inhale air from the suction hole to make the rubber membrane close to the inner wall of the membrane bearing cylinder. It is then placed outside the specimen. The gas is released. Both ends of the rubber film are turned up. The membrane carrier was taken out. The rubber band is used and the rubber membrane is fixed on the base of the pressure chamber.

(3) The sample is gently pressed with a soft brush from bottom to top to remove the bubbles between the sample and the rubber film.

(4) The drain valve is opened to allow water to flow slowly out of the sample cap to remove bubbles from the pipe. The cap is placed on the top of the sample. The bubbles are removed at the top. The rubber film is fasted on the sample cap.

(5) The water surface of the drain pipe is lowered to 20–40 cm below the elevation of the center of the sample. The excess water between the sample and the rubber film is sucked out. Then, the drain valve is closed.

(6) The pressure chamber cover is installed. During installation, the piston should be lifted to prevent collision with the sample. After the pressure chamber cover is placed, the piston should be aligned with the center of the sample cap. The screws should be tightened evenly. Then, the axial dynamometer should be aligned with the piston.

(7) The vent is opened to fill the pressure chamber with water. When the pressure chamber is almost full of water, the water inlet speed is reduced. When the water overflows from the vent, the vent is closed.

(8) The water surface of the drain pipe is flush with the height of the center of the sample. The water surface reading is measured. The drain valve is closed.

(9) The water surface of the measuring tube is at the height of the center of the sample. The metering valve is opened. The sensor is recorded. The initial reading of the pore pressure gauge is noted. Then, the measuring pipe valve is closed (required for consolidated undrained shear test, not required for consolidated drained shear test).

(10) The pipe valve and pore pressure valve are closed. The ambient pressure valve is opened to apply the required ambient pressure. The reading of each dynamometer and displacement meter is adjusted (required for consolidated undrained shear test, not required for consolidated drained shear test).

(11) The back pressure to the specimen is applied. After completion, the volume transformer valve is closed. The surrounding pressure is increased. The difference between the surrounding pressure and the back pressure equal to the original selected surrounding pressure. The stable pore pressure reading and the volume transformer water surface reading are recorded, as the initial reading before consolidation.

(12) The pore pressure valve is opened. The stable pore pressure reading is measured and recorded. The initial reading of the pore pressure gauge is subtracted. That is, the initial pore pressure u of the sample under the surrounding pressure.

(13) The drain valve and start consolidation are opened. The water surface of drain pipe and the readings of pore pressure gauge is measured and recorded. During the whole test, the water surface of the drain pipe shall be placed at the height of the center of the sample. The degree of consolidation should be at least 95%.

(14) After consolidation, the drain pipe valve or body transformer valve is closed. The readings of drain pipe or body transformer pipe and pore pressure gauge are recorded. Then, the fine adjusting hand wheel is turned. When the reading of the dynamometer starts to move, it means that the piston has contacted the sample. The reading of the axial displacement meter is recorded, which is the consolidation settlement ΔH. The height hc of the sample after consolidation can be calculated. Then the reading of dynamometer and vertical displacement meter are adjusted to zero.

6.4 Test Scheme

6.4.1 Triaxial Compression Test Scheme for Normally Consolidated Soil

The triaxial compression test of normally consolidated soil is divided into four groups. After loading, the sample is saturated by back pressure. After the specimens are completely saturated, each group is consolidated under the corresponding surrounding pressure. The completion of consolidation is based on more than 95% dissipation of pore water pressure and stable displacement. After consolidation, start the testing machine to shear the sample. The triaxial compression test scheme of normally consolidated soil under four stress paths is shown in Table 6.2. The four effective stress paths in p-q coordinate system are shown in Fig. 6.1.

6.4.2 Conventional Triaxial Compression Test Scheme for Drained Shear of Over Consolidated Soil

The same remolded clay sample is used as the over consolidation sample. After loading, the sample is saturated with back pressure. After the specimen is fully saturated, it is consolidated under a higher ambient pressure σ_c. After unloading to a lower confining pressure σ_3, it absorbs water and expands until it is stable, forming over consolidated specimen. The preparation scheme of over consolidated specimen is shown in Table 6.3.

After the water swelling of the sample is stable, the testing machine is started to shear the sample. The confining pressure (DCTC stress path) remains unchanged during the shear process. The shear strain rate is 0.014%/min. In the process of shearing, the reading of the dynamometer and the axial displacement meter is recorded. And at the same time the reading of the drainage pipe is recorded. The test was finished when the axial strain was more than 25%. The shearing process lasted about 30 h. Because the shear strain rate is a little higher, double drainage is used.

6.4.3 Elastic Deformation Parameter Test Scheme

1. **Elastic volume mod determination**

In this book, elastic bulk modulus K of soil is defined as the ratio of average normal stress to elastic strain of soil under three-dimensional equal pressure [1–3]. In order to obtain K, the isotropic consolidation test is carried out. It is assumed that pure shear does not cause elastic strain, and the change of elastic strain $\Delta\varepsilon_v^e$ is completely caused by the increment of average normal stress Δp. When the confining pressure is increased to 200, 300, 400 and 600 kPa, the consolidation is carried out. After the consolidation is completed, the samples are unloaded to 50 kPa respectively. After the free expansion of the samples is stable, the samples are compressed again. Finally, the ε_v–lg p curve (ε_v is the volume strain) is obtained at 800 kPa. Assuming that the rebound line is a straight line, the following expression can be obtained.

$$\varepsilon_v = \varepsilon_{vl} + k_v \log_{10} p \tag{6.1}$$

Table 6.2 Triaxial compression test scheme of normally consolidated soil under four stress paths

Stress path	*p*-increasing drainage shear	Equal *p* drainage shear	Conventional triaxial undrained shear	Reduced *p* drainage shear
Full name of test	Conventional triaxial compression test of consolidated drained shear	Consolidated drained shear triaxial compression test	Conventional triaxial compression test of consolidated undrained shear	Triaxial compression test of consolidated drained shear reducing confining pressure
Abbreviation	DCTC	DPTC	UCTC	DRTC
Number of samples	8	4	4	4
Consolidation pressure (kPa)	50/100/150/200/250/300/350/400	100/200/300/400	100/200/300/400	100/200/300/400
Drainage conditions	Double sided drainage	Double sided drainage	Undrained	Double sided drainage
Shear strain rate (%/min)	0.014	0.014	0.014	0.014
Stress change during shearing	Axial pressure σ_1 increasing, confining pressure σ_3 remains unchanged, and effective mean normal stress *p* increasing	Axial pressure σ_1 increases, confining pressure σ_3 decreases and effective mean normal stress *p* remains unchanged	The results show that the axial pressure σ_1 increases, the surrounding rock σ_3 remains unchanged, and the effective average normal stress *p* increases first and then decreases	The axial pressure σ_1 remains unchanged, and the effective mean normal stress *p* decreases with the decrease of girth σ_3

(continued)

Table 6.2 (continued)

Stress path	p-increasing drainage shear	Equal p drainage shear	Conventional triaxial undrained shear	Reduced p drainage shear
Data of shearing process	Reading of dynamometer, axial displacement meter and drain pipe	Reading of dynamometer, axial displacement meter and drain pipe	Dynamometer, axial displacement meter, pore pressure reading	Reading of dynamometer, axial displacement meter and drain pipe
Stress path	p-increasing drainage shear	Equal p drainage shear	Conventional triaxial undrained shear	Reduced p drainage shear
Shear completion criteria	Axial strain up to 20%	Axial strain up to 20%	Axial strain up to 20%	Axial strain up to 20%
Shear duration of specimen (h)	24	24	5	24

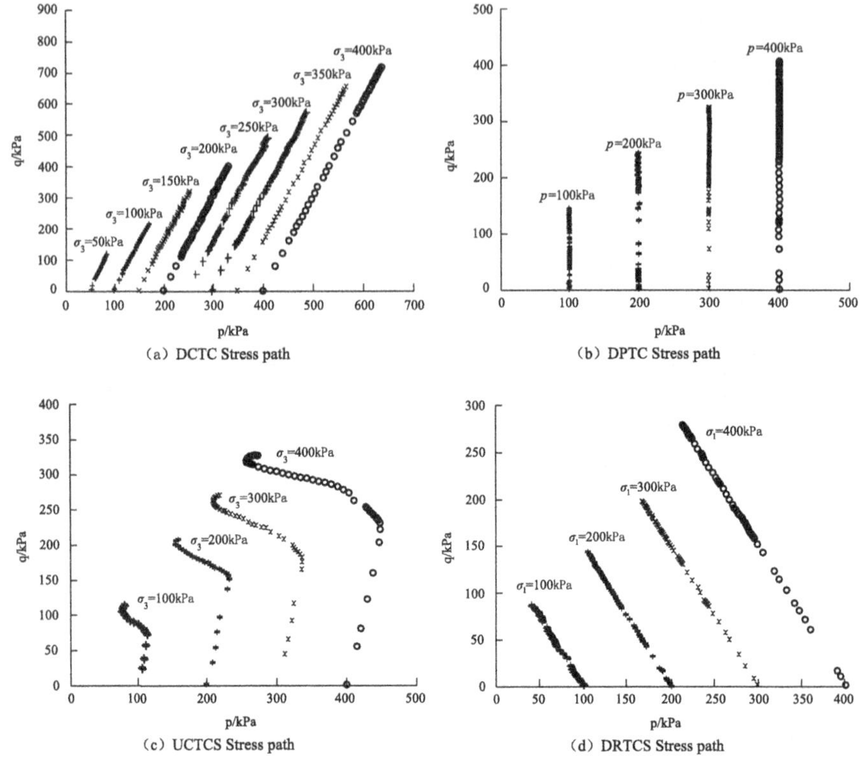

Fig. 6.1 Effective stress path of normally consolidated soil

Table 6.3 Ultra-solid sample preparation scheme

The sample number	Early solidification pressure σ_c (kPa)	After unloading the pressure σ_3 (kPa)	OCR
1	400	50	8
2	400	100	4
3	600	50	12
4	800	50	16
5	800	100	8
6	1200	150	8

where, ε_{v1} is the volume strain corresponding to p_a on the rebound line. The p_a is the atmospheric pressure ($p_a = 1.3$ kPa). k_v is the slope of rebound line, $d\varepsilon v = d\varepsilon_v^e$, $d\varepsilon_v^e$=0.43dp/p. The bulk modulus of elasticity K is as follows.

$$K = \frac{dp}{d\varepsilon_v^e} = \frac{p}{0.434k_v} \tag{6.2}$$

From the ε_v–lg p curve, the corresponding K value of spring back is obtained, when p is 200 kPa, 300 kPa, 400 kPa and 600 kPa, respectively. The relationship between K/P_a and P/P_a is drawn on a double logarithmic curve, which is approximately a straight line and can be written as

$$K = K_1 p_a \left(\frac{p}{p_a}\right)^n \tag{6.3}$$

where, K_1 and n are test constants, K_1 is P_a The corresponding K value. K value and n value are obtained by linear regression.

2. **Determination of elastic shear modulus**

In this book, elastic shear modulus G of soil is defined as the ratio of generalized shear stress to elastic shear strain [4–6]. In order to obtain G, the conventional triaxial compression tests of consolidated drained shear with loading and unloading cycles at σ_3 of 50, 200 and 400 kPa are carried out, and the q curve is calculated by $\varepsilon s = \varepsilon 1 - \varepsilon v/3$ ($\varepsilon s$is shear strain). The springback slope is different under different σ_3. If we neglect this effect, then $G(\sigma 3) = \mathrm{d}p/\mathrm{d}\varepsilon_s^e$, if we plot it on $\lg(G/p_a) - \lg(\sigma_3/p_a)$ graph and assume that it is a straight line, then

$$\lg G = \lg(G_1 p_a) + n \lg\left(\frac{\sigma_s}{p_a}\right) \tag{6.4}$$

where, G_1 and n are test constants.

The following formula is obtained from Eq. (6.4).

$$G = G_1 p_a \left(\frac{\sigma_s}{p_a}\right)^n \tag{6.5}$$

In general triaxial compression test, $\sigma 3 = p - \frac{1}{3}(\sigma 1 - \sigma 3) = p - \frac{1}{3}q$, so there is

$$G = G_1 p_a \left(\frac{p - \frac{1}{3}q}{p_a}\right)^n \tag{6.6}$$

The curve of $\lg(G/p_a)$–$\lg(\sigma_3/p_a)$ is obtained by averaging the rebound line and the recompression line. The values of G_1 and n are obtained by linear regression.

6.5 Test Results and Analysis

6.5.1 Results and Analysis of the Three-Axis Compression Test of Normal Solidified Soil

The relationships between $\sigma_1-\sigma_3-\varepsilon_1$ and $\varepsilon_v-\varepsilon_1$ of normally consolidated soil samples under four stress paths are shown in Fig. 6.2 and Fig. 6.3 respectively. It can be seen that the remolded clay samples show strain strengthening characteristics under the four stress paths. In the initial stage, the principal stress difference increases with the increasing rate of axial strain (its stiffness) from DCTC-DPTC-UCTC-DRTC. After yielding, the growth rate of principal stress difference with axial strain (its stiffness) gradually flattens from stress path DCTC-DPTC-UCTC-DRTC. In view of the strain strengthened specimen, the principal stress difference corresponding to the axial strain of 15% is generally taken as its strength value [7, 8]. It can also be seen from the Figs. 6.2 and 6.3 that the strength value of the sample is the largest under the increasing p drainage stress path. The strength under the equal p drainage stress path is about 1/2 of that. The strength of conventional triaxial undrained stress path is about 1/2 of that. The strength value under the path of reducing p drainage stress is

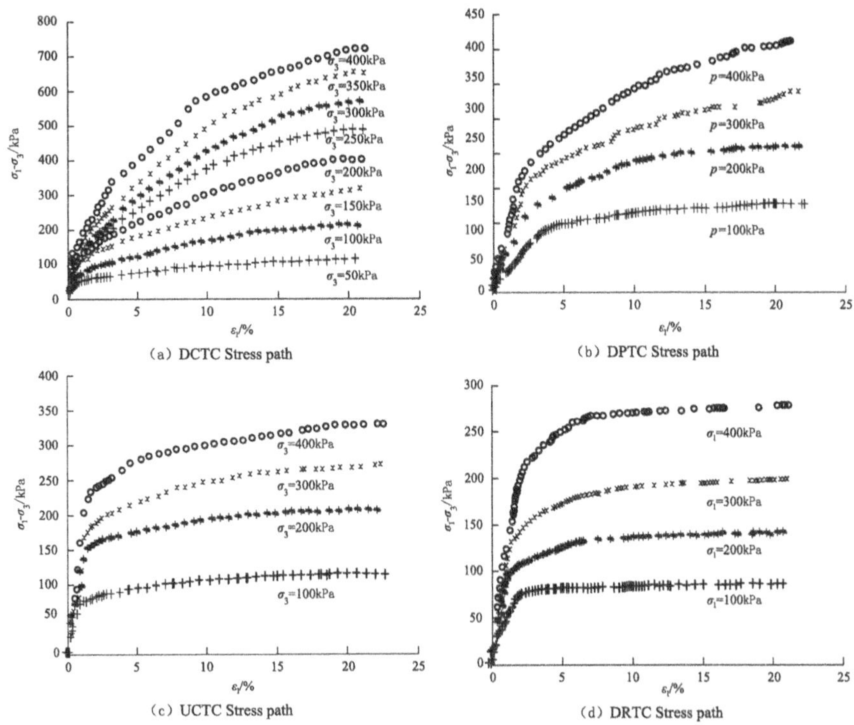

Fig. 6.2 Normal solidified soil $\sigma_1-\sigma_3-\varepsilon_1$ relationship

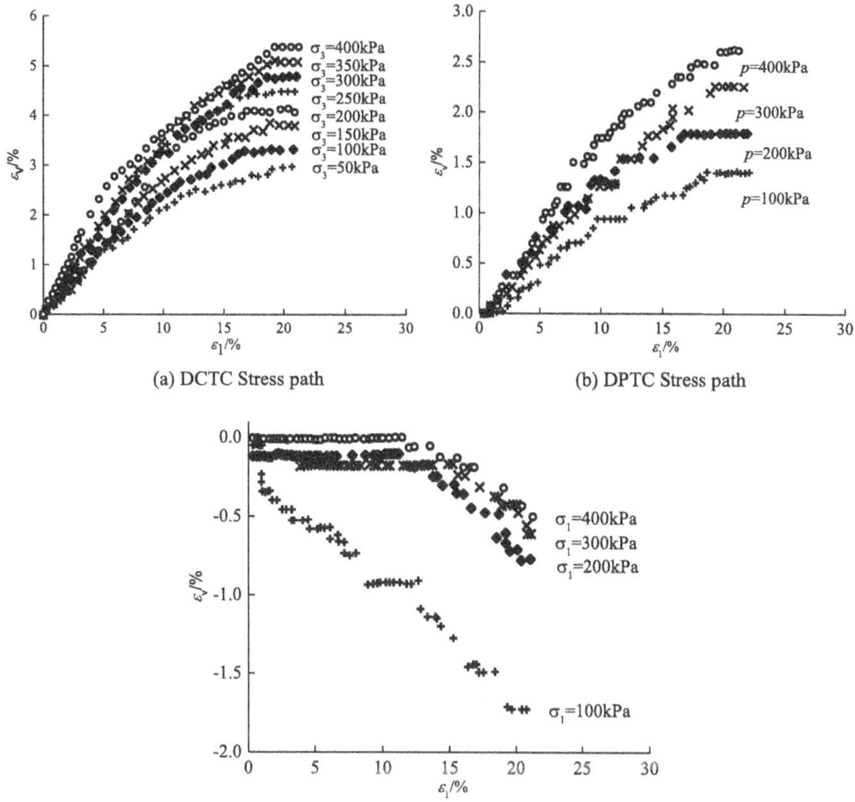

(c) DRTC Stress path

Fig. 6.3 Relationship between ε_v–ε_1 of normally consolidated soil

about 1/3 of that. The results show that the strength of the specimen decreases from DCTC-DPTC-UCTC-DRTC.

For the volume strain, it shows pure shear shrinkage under increasing p and equal p drainage stress paths. The volume shrinkage increases with the increase of consolidation pressure. The volume strain amplitude under the increasing p stress path is about twice that under the corresponding equal p-stress path. However, the volumetric strain shows pure dilatancy under the p-stress path, and the volumetric expansion decreases with the increase of consolidation pressure. That is to say, the specimen with higher consolidation pressure has less expansion.

The above phenomena show that the effect of stress path on the stiffness, strength and volume deformation characteristics of normally consolidated soil is considerable and can not be ignored [9, 10].

6.5.2 Results and Analysis of Three-Axis Compression Test of Ultra-Solidified Soil

For the convenience of comparative analysis, the stress–strain curves of over consolidated soil samples and corresponding normally consolidated soil samples are drawn together [11, 12]. Figures 6.4, 6.5, 6.6, 6.7, 6.8 and 6.9 show the comparison of stress–strain curves of clay samples under different consolidation conditions. Figure 6.4 shows the case of OCR = 8. Figure 6.5, Fig. 6.6 and Fig. 6.7 show the shear confining pressures of 50 kPa, 100 kPa and 150 kPa respectively under different OCR conditions. Figure 6.8 and Fig. 6.9 show the conditions under which the pre consolidation pressure σ_c is 400 kPa and 800 kPa respectively and OCR is different.

Generally speaking, the normally consolidated soils show the characteristics of strain strengthening and shear shrinkage, while the over consolidated soils show the characteristics of strain softening and shear dilatancy.

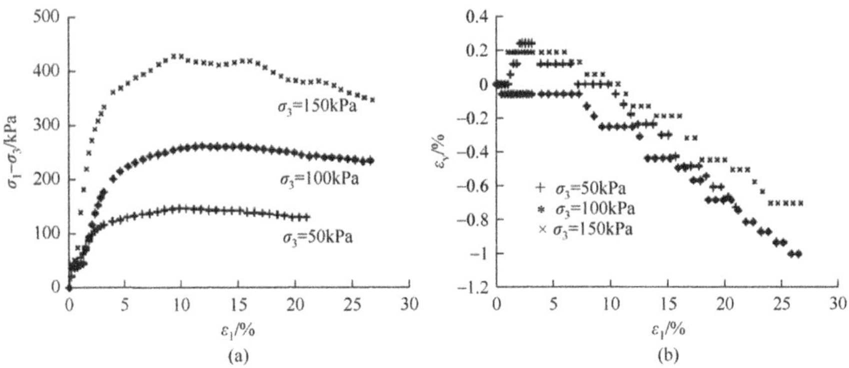

Fig. 6.4 Stress–strain relationships under the same OCR (OCR = 8)

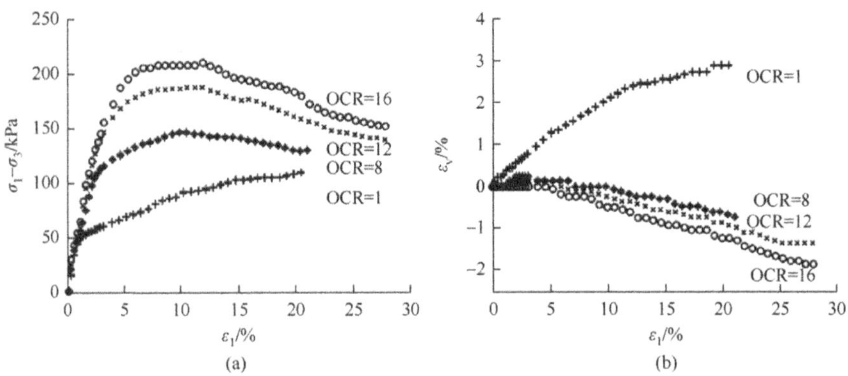

Fig. 6.5 Stress strain relationship under different OCR (σ_3 = 50 kPa)

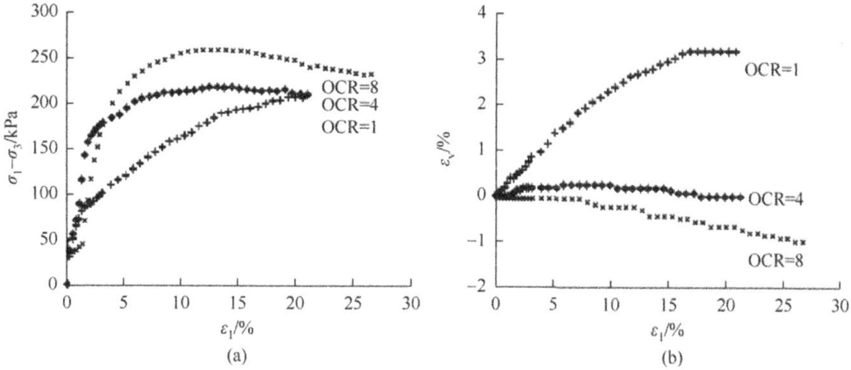

Fig. 6.6 Stress strain relationship under different OCR ($\sigma_3 = 100$ kPa)

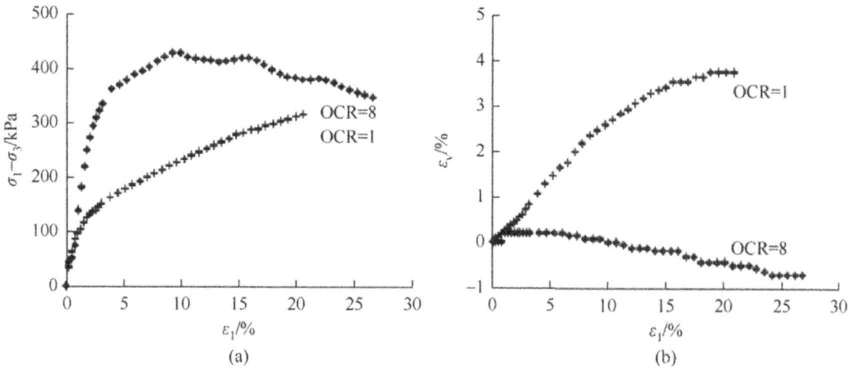

Fig. 6.7 Stress strain relationship under different OCR ($\sigma_3 = 150$ kPa)

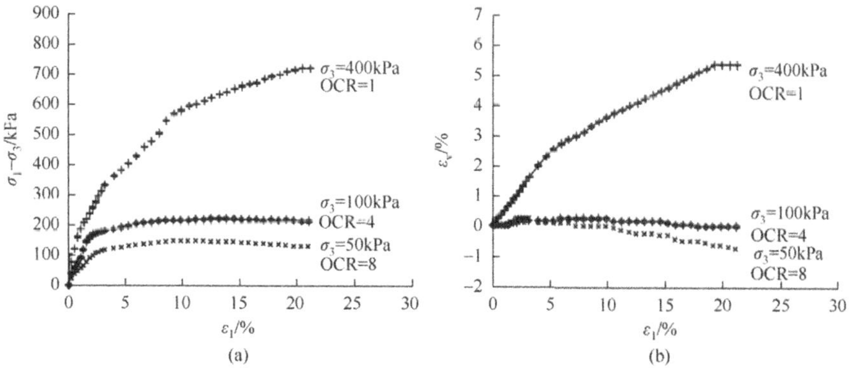

Fig. 6.8 Stress strain relationship under different OCR ($\sigma_3 = 400$ kPa)

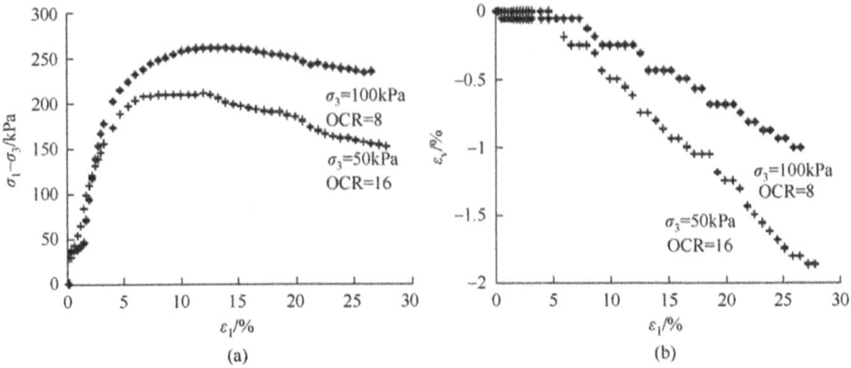

Fig. 6.9 Stress strain relationship under different OCR ($\sigma_3 = 800$ kPa)

It can be seen from Fig. 6.4 that under the condition of fixed OCR, the volume strain of the specimen is not sensitive to different consolidation pressures (including pre consolidation pressure σ_c and consolidation pressure σ_3 during shear), and its shear resistance is closely related to the consolidation pressure.

It can be seen from Figs. 6.5, 6.6 and 6.7 that the shear resistance of the specimen increases with the increase of OCR under the condition of fixed consolidation pressure σ_3. With the increase of OCR, the hump phenomenon becomes more and more obvious. The shear strength of specimens with different OCR tends to a stable value. That is, the strength is on the critical state line. For the volume strain, with the increase of OCR, the specimen changes from shear shrinkage (normally consolidated soil) to shear dilatation (over consolidated soil). The larger the OCR is, the more obvious the dilatancy is.

It can be seen from Figs. 6.8 and 6.9 that the shear strength is related to the consolidation pressure and OCR under the premise of fixed pre-consolidation pressure σ_c. For the volume strain, OCR is still the decisive factor, that is, the specimen with larger OCR has larger dilatancy.

In conclusion, for volume strain, OCR is the decisive factor. The volume strain is not sensitive to consolidation pressure. For the shear capacity, the consolidation pressure is the decisive factor, and the influence of OCR can not be ignored. OCR determines whether it is strain hardening or strain softening. It also determines the degree of strain softening. However, the specimen will eventually reach a unified critical state with approximately the same residual strength.

The above conclusion is obtained by comparing the stress–strain relationship of normally consolidated soil and over consolidated soil under the stress path of conventional triaxial compression test (DCTC stress path). This conclusion does not consider the influence of other stress paths.

6.5.3 Test Results of Elastic Deformation Parameters

In order to obtain the elastic bulk modulus K, the isotropic isobaric consolidation test is carried out [13, 14]. The σ_3 is added to consolidate at 200 kPa, 300 kPa, 400 kPa and 600 kPa, respectively. After the consolidation is completed, it is unloaded to 50 kPa respectively. After the free expansion and stability of the sample, it is compressed again respectively, and finally to 800 kPa. The ε_v–lg p relationship is shown in Fig. 6.10. In order to obtain the elastic shear modulus G, conventional triaxial compression tests of consolidated drained shear with loading and unloading cycles under confining pressures of 50, 200 and 400 kPa are carried out. The relationship between σ_1–σ_3–ε_1 is shown in Fig. 6.11. Some specimens after failure are shown in Fig. 6.12.

Thus, the elastic bulk modulus K of the sample is measured, as shown in Eq. (6.7). The elastic shear modulus G of the sample is measured, as shown in Eq. (6.8).

$$K = 158.161 p_a \left(\frac{p}{p_a} \right) 0.7301 \tag{6.7}$$

$$G = 327.11 p_a \left(\frac{p - q/3}{p_a} \right) 0.8885 \tag{6.8}$$

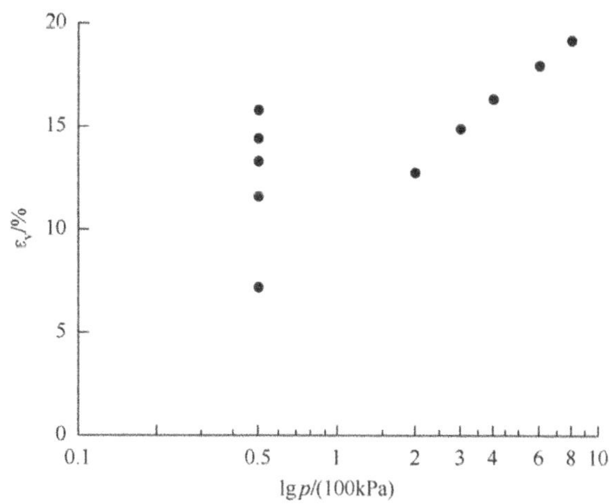

Fig. 6.10 Relationship of ε_v–lg p

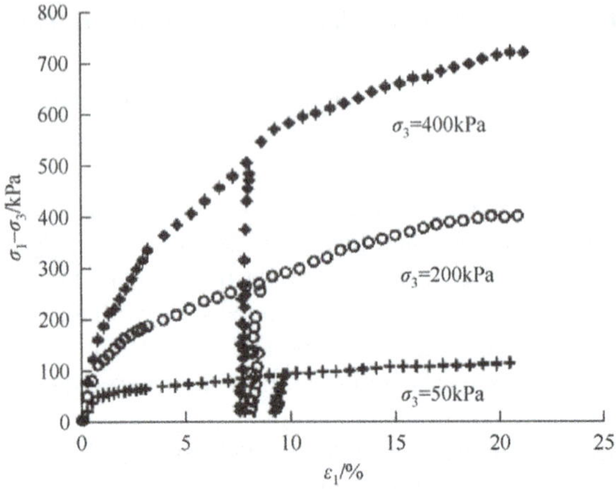

Fig. 6.11 Relationship of $\sigma_1-\sigma_3-\varepsilon_1$

Fig. 6.12 Partial specimens after failure

6.5.4 Analysis of Test Results

By comparing the stress–strain curves of remolded clay under different stress paths and stress histories, the following conclusions can be drawn.

Under the four stress paths, the normally consolidated soil samples show strain strengthening characteristics. In the initial stage, the principal stress difference increases with the increasing rate of axial strain (its stiffness) from DCTC-DPTC-UCTC-DRTC. After yielding, the growth rate of principal stress difference with axial strain (its stiffness) gradually flattens from stress path DCTC-DPTC-UCTC-DRTC. The results show that the strength of the specimen is the largest under the increasing *p* drainage stress path. The strength under the equal *p* drainage stress path is about 1/2 of that. The strength of conventional triaxial undrained stress path is about 1/2 of that. The strength value under the path of reducing *p* drainage stress is about 1/3 of that. That is to say, the strength of the specimen decreases from DCTC-DPTC-UCTC-DRTC. For the volume strain, it shows pure shear shrinkage under both increasing *p* and equal *p* stress paths, and its volume shrinkage increases with the increase of consolidation pressure [15–17]. The volume strain amplitude under the increasing *p* stress path is about twice that under the corresponding equal *p* stress path. However, the volumetric strain shows pure dilatancy under the *p*-stress path, and the volumetric expansion decreases with the increase of consolidation pressure. That is to say, the specimen with higher consolidation pressure has less expansion. These phenomena show that the effect of stress path on the stiffness, strength and volume deformation characteristics of normally consolidated soil is considerable and cannot be ignored.

In the aspect of the influence of stress history on the constitutive relationship of soil, the over consolidation ratio is the decisive factor for the volume strain, while the volume strain is not sensitive to the consolidation pressure. For the shear capacity, the consolidation pressure is the decisive factor. The effect of over consolidation ratio can not be ignored. The over consolidation ratio determines the strain hardening or strain softening, and the degree of strain softening. However, the specimen will eventually reach a unified critical state with approximately the same residual strength.

6.6 Numerical Modeling of Clay Constitutive Relation

With the expansion of soil engineering and scale, the establishment of soil elasto-plastic constitutive model which can reflect the stress path correlation has become an important problem to be solved. Similar to Chaps. 4 and 5, this chapter attempts to extend the numerical modeling method of geotechnical constitutive relationship to the field of cohesive soil. Based on the triaxial compression tests of normally consolidated soil under the stress paths of increasing *p* drained shear, equal *p* drained shear, conventional undrained triaxial shear and decreasing *p* drained shear, the elastic–plastic constitutive models of clay under these four stress paths are established.

Compared with the classical constitutive model, the numerical modeling method shows the advantages.

1. **Yield surface analysis under four stress paths**

In this section, the double yield surface theory is used to establish the yield surface. The total strain is measured by triaxial compression test, and the plastic strain is separated from the total strain by elastic bulk modulus and elastic shear modulus. Then, the plastic shear strain and the plastic volumetric strain (where the plastic volumetric strain takes the consolidation starting point as the starting point) are taken as hardening parameters respectively. By drawing its contour on *p*-*q* plane, the trajectories of shear and volume yield can be obtained. Based on this, the shear yield trajectories (Fig. 6.13) and volume yield trajectories (Fig. 6.14) under four stress paths are plotted.

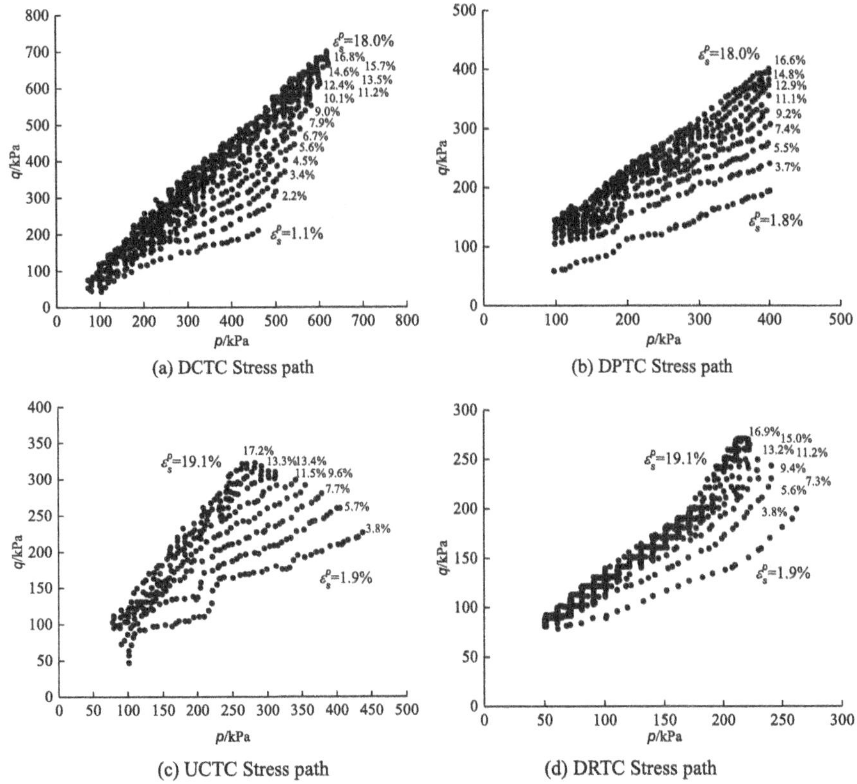

Fig. 6.13 Shear yield trajectory of normally consolidated soil under four stress paths

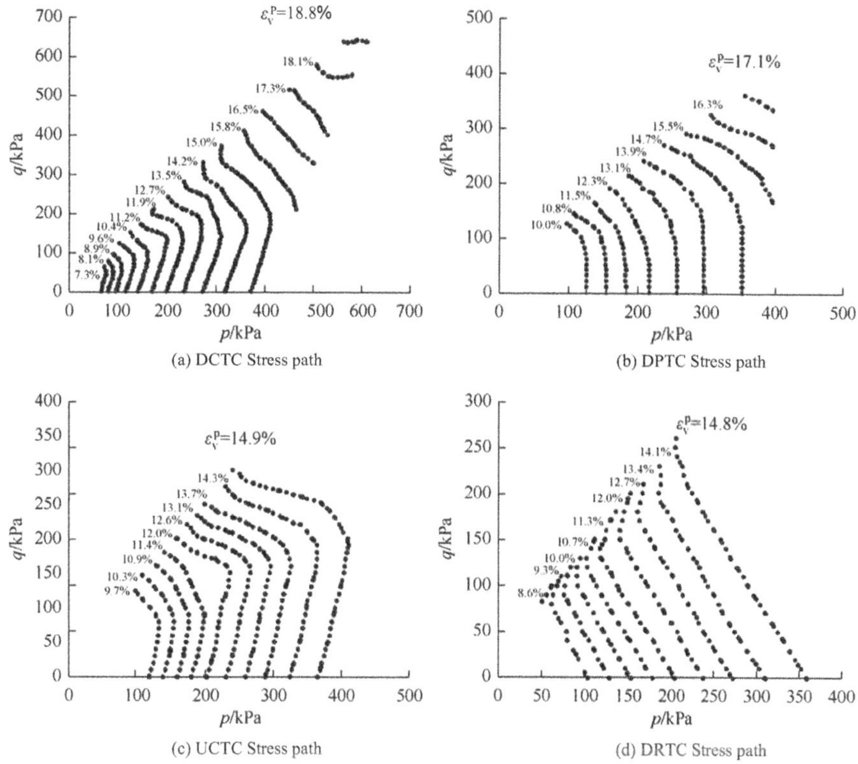

Fig. 6.14 Volume yield trajectory of normally consolidated soil under four stress paths

It can be seen from Figs. 6.13 and 6.14 that the basic change trend of shear yield trajectory of specimens under four stress paths is relatively consistent. However, from the stress path DCTC-DPTC-UCTC-DRTC, the rising rate of plastic shear strain increases gradually. For the volume yield trajectory, it begins to deflect slightly to the right and then slowly to the left under the DCTC stress path. It deflects monotonously to the left under the DPTC stress path. Under the stress path of UCTC, it deflects slightly to the right at the beginning and then to the left rapidly. This shows that the volume yield trajectory is not different from the effective stress path trajectory. This is because along the undrained effective stress path, the total volume of the specimen remains unchanged. The increase of the elastic strain is equal to the decrease of the plastic strain. However, this change is small, so the volume yield trajectory depends largely on the plastic volumetric strain after consolidation. At the beginning of the DRTC stress path, it deflects monotonously to the left, and near the failure, it dilates and deflects to the right.

Therefore, the following important conclusions can be drawn. The comparison of shear and volume yield trajectories of normally consolidated soils under different stress paths shows that the stress path has a significant effect on the evolution of strain hardening of clay. In the traditional modeling method of soil elastoplastic constitutive model, it is inaccurate to describe the stress–strain relationship of soil by assuming the fixed form of yield surface. As shown in Figs. 6.13 and 6.14, the yield surface of soil under different stress paths is significantly different.

2. **Establishment of elastic–plastic constitutive model of clay**

Based on the basic framework of numerical modeling method (see Chap. 3), the triaxial test data sets (p, q, ε_{V}, ε_{S}) of clay under four stress paths are fitted by Gauss method to obtain the stress–strain relationship of the whole stress field (p, q). The relationship surface of ε_{S}-(p, q) under four stress paths is shown in Fig. 6.15, and the relationship surface of ε ε_{V}-(p, q) is shown in Fig. 6.16.

It can be seen from Fig. 6.15 that the ascending slope of the specimen's ε_{S}-(p, q) surface under the four stress paths is different. The stress path from DCTC-DPTC-UCTC-DRTC is steeper. This is consistent with the change of shear yield trajectory. For the ε_{V}-(p, q) surface, the peak value under DCTC stress path is higher than that under DPTC stress path. The volume strain of DRTC is negative. This means that the

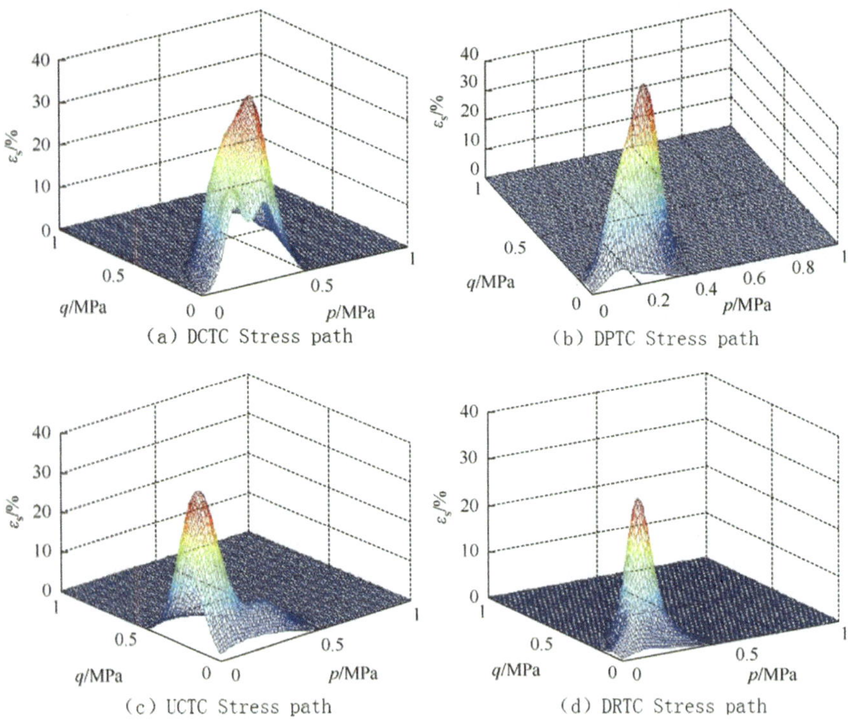

Fig. 6.15 Relationship of ε_{S}-(p, q) for normally consolidated soil under four stress paths

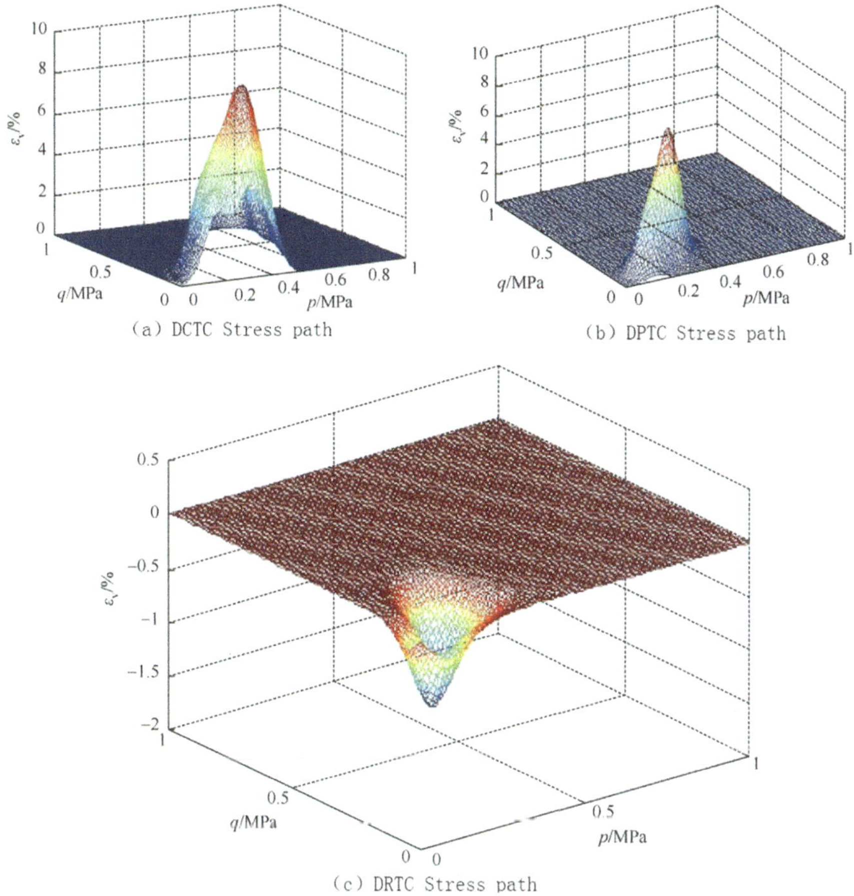

Fig. 6.16 Relationship of ε_V-(p, q) for normally consolidated soil under three stress paths

whole deformation process is volume expansion. It can also be seen from Figs. 6.15 and 6.16 that the stress range of the stress–strain curve surface is quite different under different stress paths. This is consistent with the range of four effective stress paths in the p-q coordinate system shown in Fig. 6.1.

Figures 6.15 and 6.16 and their comparative analysis show that the numerical modeling method can comprehensively describe the deformation characteristics of soil under different stress paths.

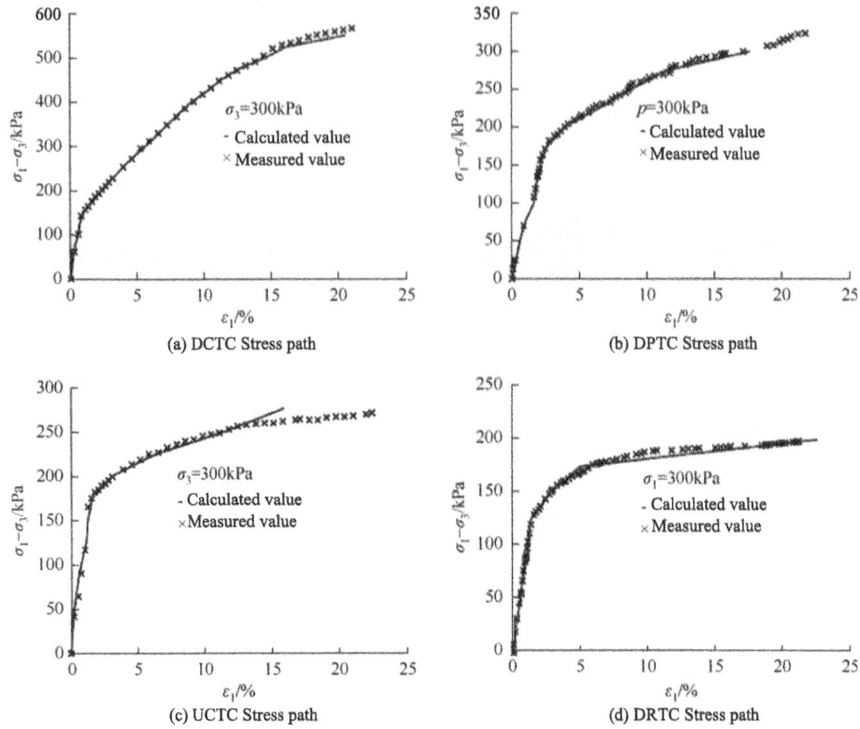

Fig. 6.17 Relationship between theoretical prediction and measured σ_1–σ_3–ε_1

3. **Verification of numerical constitutive model**

In order to verify the correctness of the numerical constitutive model, the corresponding stress–strain calculation programs are compiled for the four loading stress paths [18–20]. Figure 6.17 shows the comparison between the calculated value and the measured value of the relationship ofσ_1–σ_3–ε_1 under the four stress paths, and Fig. 6.18 shows the comparison between the calculated value and the measured value of the relationship of ε_v–ε_1. It can be seen from the Fig. 6.18, the model prediction is in good agreement with the experimental results. The numerical modeling method can better reflect the axial deformation and volume deformation characteristics of soil in a large strain range [19–23].

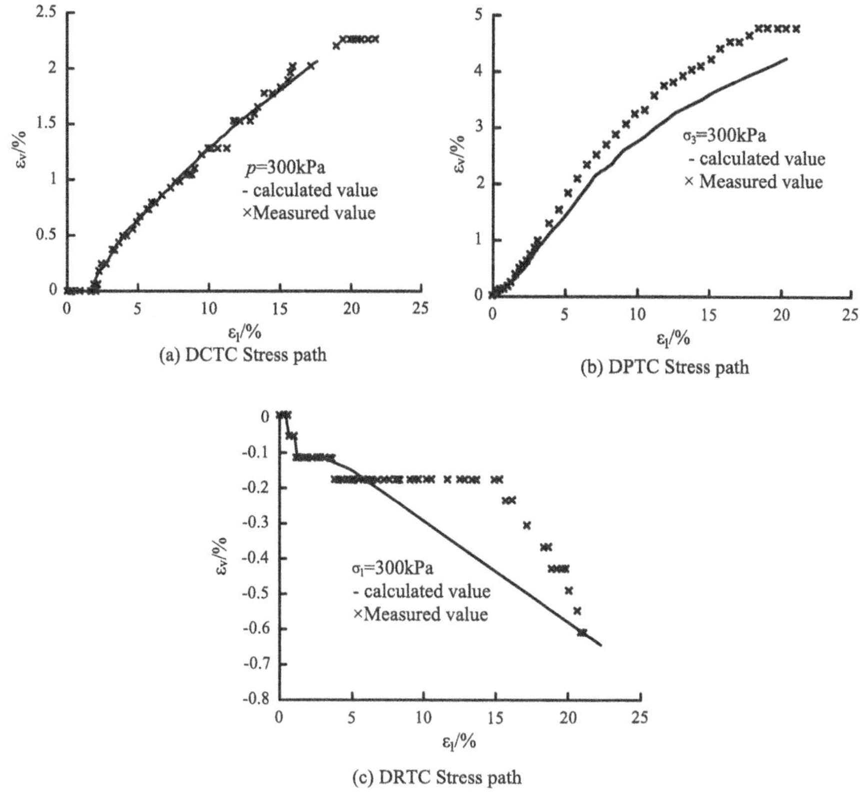

Fig. 6.18 Relationship between theoretical prediction and measured ε_v–ε_1

References

1. Cheng T, Yan K, Wang J (2009) Study of numerical model for triaxial compression test on clay under different consolidation conditions. Rock Soil Mech 30(11):3352–3356
2. Kong L, Zhou B, Bai H et al (2010) Experimental study of deformation and strength characteristics of Jingmen unsaturated expansive soil. Rock Soil Mech 31(10):3036–3042
3. Li D, Wang R, Hu P (2007) Creep damage coupled constitutive relation of frozen clay. Glacial Frozen Soil 29(3):446–449
4. Ren Q, Wang J (2005) Triaxial experimental studies of expansive soil's constitutive relationship with various moisture contents. Soil Eng Found 19(5):55–57
5. Ren Q, Wang J (2005) Numerical method in modeling the elastoplastic constitutive relationship of sand under the path of constant proportion of principle stress. J Huazhong Univ Sci Technol (Urban Sci) 22(3):69–71
6. Sun L (2014) Experimental study on statistical damage constitutive model of unsaturated soil and its engineering application. Hunan University, Changsha
7. Wang J (2002) Numerical method in modeling the constitutive relations of rock and soil. J Huazhong Univ Sci Technol (Urban Sci) 19(1):44–47
8. Xiao J, Yang H, Li H et al (2013) Shear strength test of Nanning expansive soil with various dry densities and low stresses. China J Highw Transp 26(6):15–21

9. Xin Z, Tan X, Hu N et al (2014) Experimental study and variability analysis of swell-shrinkage indices of expansive soil. J Guangxi Univ (Nat Sci Ed) 39(1):124–131
10. Xu B, Yin Z, Liu S (2011) Experimental study of factors influencing expansive soil strength. Rock Soil Mech 32(1):44–50
11. Yao H, Wu Y, Huang B (2010) Experimental study on swelling characteristics of expansive soil under lateral restricted conditions. Water Resour Power 28(7):40–43
12. Zhang P, Liu D, Huang D et al (2003) Saturated-unsaturated unsteady seepage flow numerical simulation. Rock Soil Mech 24(6):927–930
13. Katti DR, Desai CS (1996) Modeling and testing of cohesive soil using disturbed-state concept. Int J Rock Mech Min Sci Geomech Abstr 33(3):114–130
14. Li X, Thomas HR, Fan Y (1999) Finite element method and constitutive modelling and computation for unsaturated soils. Comput Method Appl Mech Eng 169(1/2):356–367
15. Li X, Thomas HR, Fan Y (1999) Finite element method and constitutive modelling and computation for unsaturated soils. Comput Methods Appl Mech Eng 169(1/2):135–159
16. Norris VA (1982) Numerical modeling of soil response to cyclic loading "stress-reversal surfaces". In: International symposium numerical models in geomechanics, Zurich, pp 38–49
17. Oettl G, Stark RF, Hofstetter G (1998) A comparison of elastic-plastic soil models for 2D FE analyses of tunneling. Comput Geotech 23(1/2):19–38
18. Ren QY, Wang JT (2005) Numerical method in modeling the elastoplastic constitutive relationship of sand under different paths. J China Geosci 16(3):268–270
19. Shin HS, Pande GN (2000) On self-learning finite element codes based on monitored response of structure. Comput Geotech 27:161–178
20. Simpson B, Riordan NJ, Croft DD (1979) A computer model for the analysis of ground movements in London clay. Geotechnique 29(2):149–173
21. Sun DA, Matsuoka H, Yao YP et al (2004) An anisotropic hardening elastoplastic model for clays and sands and its application to FE analysis. Comput Geotech 31(1):37–46
22. Woodward PK, Berenji AP (2001) Advanced numerical investigation of Terzaghi's superposition theory. Adv Eng Softw 32(10):797–804
23. Zhang HW, Sanavia L, Schrefler BA (2001) Numerical analysis of dynamic strain localisation in initially water saturated dense sand with a modified generalised plasticity model. Comput Struct 79:441–459

Chapter 7
Influence of Stress Path and Stress History on the Constitutive Relation

The influence of stress history and stress path on soil constitutive relationship is well known. However, the mechanism and degree of its influence are not very clear. This chapter is based on the principle of interaction between plastic volumetric strain and plastic shear strain. Based on the triaxial compression tests of remolded clay under different stress paths and stress histories, the influence mechanism of stress history and stress path on the constitutive relationship of clay is clarified. In addition, in order to explore the effect of stress path on the effective shear strength parameters of clay. Under the premise of ensuring that the initial state, stress history, drainage conditions, loading rate, test instruments and failure criteria of the samples are consistent, the triaxial compression tests of the same normally consolidated remolded clay under different stress paths are carried out. The test results show that the effective shear strength parameters of remolded clay under different stress paths are quite different, and this is qualitatively analyzed.

7.1 Mechanism of the Effect of Stress Path on Soil Constitutive Relation

In metal plastic deformation, the stress–strain relationship is related to the stress path, which is manifested in loading, unloading and reloading [1–3]. However, in the plastic deformation of soil, a large number of triaxial tests and engineering practice have confirmed that the stress–strain relationship is not only related to the stress path in loading, unloading and reloading. Moreover, the stress path dependence is obvious under monotonic loading. In soil engineering, the main stress paths of soil are quite different in some cases. Especially, for underground engineering, slope and deep foundation pit engineering, the surrounding soil is in the stress state of decreasing confining pressure in the process of excavation, which is the so-called stress reduction path. Compared with the increasing p stress path of the foundation, it has directional difference. In addition, the constitutive relationship of soil under

Q. Ren and J. Zhou, *Numerical Modeling of Soil Constitutive Relationship*,
https://doi.org/10.1007/978-981-16-3231-0_7

cyclic loading and large stress transition is one of the hot topics in soil mechanics. In fact, cyclic loading or large stress transition is a special stress path. Under these stress paths, the yield surface rotates, showing the characteristics of rotational hardening.

However, what is the mechanism of stress path dependence of soil deformation? At present, there are different explanations [4, 5]. Some researchers think it is caused by anisotropy, and some attribute it to the cross action of p and q. In this section, based on the triaxial compression test results of clay under different stress paths, according to the principle of ε_v^p interaction, it is proved that the stress path correlation is the comprehensive expression of ε_v^p interaction. In addition, the mechanism of rotational hardening under cyclic loading and large stress transition is analyzed, and it is clearly pointed out that the root of yield surface rotation is stress path dependence.

7.1.1 Test Work

In order to explore the mechanism of stress path dependence of soil deformation, consolidated drained shear triaxial compression tests of clay under different stress paths are carried out. The preparation, saturation, installation and consolidation process of clay samples are shown in Sects. 6.1–6.3, and the physical properties of samples are shown in Table 6.1.

The consolidated drained shear triaxial compression test is carried out in three groups. In the shear process, the first group of specimens adopts one of the increasing p stress paths (DCTC path). In the second group, the equal p stress path (DPTC path) is used. In the third group, one of the p-stress reduction paths (DRTC path) is used. The triaxial compression test scheme of normally consolidated soil under three stress paths is shown in Table 6.2. The three effective stress paths in p-q coordinate system are shown in Fig. 6.1.

The isolines of plastic volumetric strain and plastic shear strain are taken as volume and shear yield locus, respectively. The volume and shear yield trajectories under the above three stress paths are plotted by the method of drawing yield surface directly using triaxial test results proposed by Huang Wenxi, which are shown in Figs. 6.13 and 6.14.

7.1.2 Stress Path Dependence Is a Comprehensive Expression of the Interaction Between Plastic Bulk Strain and Shear Strain

According to the principle of $\varepsilon_v^p - \varepsilon_s^p$ interaction, the average normal stress p does not directly act on the plastic shear strain [6, 7]. It is the effect of plastic volumetric strain on plastic shear strain. Similarly, the effect of generalized shear stress q on plastic volumetric strain is not direct. It is realized by plastic shear strain. The details

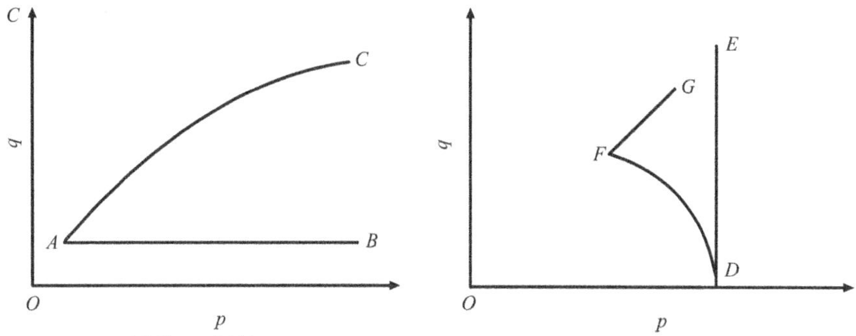

(a) Shear yield locus (b) Volume yield locus

Fig. 7.1 Shear and volume yield trajectories with or without ε_v^p interaction

of the $\varepsilon_v^p - \varepsilon_s^p$ interaction is shown in Fig. 3.1. The change of plastic volumetric strain leads to the rise and fall of shear resistance, which affects the plastic shear strain. The plastic shear strain acts on the volume strain through dilatancy or shrinkage. In addition to direct compression, the average normal stress can restrain the dilatancy.

It can be seen from the above analysis that the effects of p and q on the plastic shear strain and plastic volume strain respectively actually reflect the interaction of $\varepsilon_v^p - \varepsilon_s^p$. It can be directly observed from Figs. 6.13 and 6.14 that all shear and volume yield trajectories are not parallel to the p-axis or q-axis and are generally curves. In order to clarify the correlation between yield trajectory and stress path, shear and volume yield trajectories with or without $\varepsilon_v^p - \varepsilon_s^p$ the interaction are plotted in Fig. 7.1a, b, respectively.

If there is no $\varepsilon_v^p - \varepsilon_s^p$ interaction, the shear and volume yield trajectories must be straight lines, as shown by lines AB and DE in Fig. 7.1. If the yield trajectory is a curve, as shown by AC and DFG in Fig. 7.1, there must be $\varepsilon_v^p - \varepsilon_s^p$ interaction. According to the triaxial compression test results, as shown in Figs. 6.13 and 6.14, the trajectories of shear and volume yield under three stress paths are not parallel to the p-axis or q-axis, which are curves. This confirms the interaction between ε_v^p and ε_s^p in the plastic deformation of clay.

It can be seen from the above analysis that the effect of stress path on yield trajectory, as shown in Fig. 7.1, includes both the effect of ε_v^p on ε_s^p and the effect of ε_s^p onε_v^p. Therefore, the correlation of stress path is the comprehensive expression of $\varepsilon_v^p - \varepsilon_s^p$ interaction.

It can also be seen from Figs. 6.13 and 6.14 that under the three stress paths, the absolute values of the average slopes of the shear and volume yield trajectories are not the same, and they gradually increase with the stress path changing order DRTC-DPTC-DCTC. At the same time, the maximum shear stress corresponding to the maximum strain is different under the three stress paths. The above two cases are the manifestation of the effect of stress path on yield surface.

7.1.3 Mechanism of Rotational Hardening

In the past few decades, the study of soil properties under cyclic loading or large stress transition has been developed rapidly [8–10]. Some elastoplastic constitutive models have been proposed, which has become one of the hot spots in soil mechanics. In metals, the induced anisotropy leads to the translation of the yield surface, which can be easily described by motion hardening. However, more and more experimental data show that under the condition of large stress transition, the soil properties change, resulting in the rotation rather than translation of the yield surface. Some researchers think that the rotation of soil yield surface is the manifestation of anisotropy.

According to the triaxial compression test results of clay under three stress paths given in Sects. 5.2.1 and 7.1.2, the results show that the plastic deformation of soil is indeed related to the stress path, and even for monotonic loading paths in different directions. There are significant differences between the corresponding yield surfaces, which are shown in two aspects. The absolute value of the average slope of the shear and volume yield trajectories changes with the change of the stress path direction. The volume and shear yield trajectories are in the stress field (p, q) The range of expansion and contraction also varies with the stress path. This fact means that when the direction of monotonic loading stress path changes, the corresponding volume and shear yield surface will rotate as a whole, accompanied by the change of yield surface size in stress space. This phenomenon is the performance of soil rotation hardening.

According to the analysis results in Sect. 5.2.2, the stress path correlation is a comprehensive expression of the interaction between plastic volumetric strain and plastic shear strain. In fact, due to the change of the direction of the stress path, the mode of interaction between the corresponding plastic volumetric strain and plastic shear strain also changes, from one monotonic path to another, resulting in the rotation of the yield surface. Because the yield surface of soil should always contain the origin in the stress space, it can only be allowed to rotate, not to translate. Both cyclic loading and large stress transition are special stress paths, which are also the comprehensive expression of the interaction between plastic volumetric strain and plastic shear strain. Therefore, the rotation of yield surface under cyclic loading and large stress transition is also caused by stress path dependence.

The above demonstration of stress path dependence and rotational hardening mechanism is another confirmation of the interaction principle between plastic bulk strain and plastic shear strain. In cyclic loading, Bauschinger effect occurs when the stress changes from compressive stress to tensile stress. For soil unloading, Li Guangxin pointed out that when the load is reduced to a certain extent, it will yield and produce plastic deformation. Based on the previous analysis, the above two cases may be the result of stress history.

Based on the triaxial compression test results of clay under different stress paths, according to the principle of $\varepsilon_v^p - \varepsilon_s^p$ interaction, it is theoretically proved that the correlation between soil plastic deformation process and stress path is the comprehensive performance of $\varepsilon_v^p - \varepsilon_s^p$ interaction.

Triaxial compression tests of clay under monotonic loading paths in different directions show that when the stress path direction changes, both the volume and shear yield surfaces rotate in the stress space, accompanied by the expansion and contraction of the stress range of the yield surface. This phenomenon can be called rotational hardening, which is actually the reflection of stress path dependence on the yield surface.

A large number of experimental evidences show that rotational hardening occurs under cyclic loading or large stress transition. Because both cyclic loading and large stress transition are special stress paths, they are composed of monotonic loading paths. Therefore, the rotation of yield surface under cyclic loading or large stress transition is still the result of stress path, which is essentially the manifestation of $\varepsilon_v^p - \varepsilon_s^p$ interaction.

7.2 Research on the Mechanism of the Influence of Stress History on the Constitutive Relationship of Clay

The elasto-plastic deformation of soil shows diversity and complexity, with nonlinearity, compression, dilatancy and stress path dependence. However, in the conventional triaxial compression test, a large number of test results show that the stress–strain curve of soil can be divided into two basic types. One is the strain strengthening type represented by the normally consolidated soil. The other is the strain softening curve of over consolidated soil with hump. In the whole shear deformation process, the volume of the former shrinks continuously. The latter is in a state of volume expansion until it reaches the critical state, except for a slight volume contraction at the initial stage.

Drucker et al. Proposed a framework of critical state soil mechanics based on their theoretical and experimental work in Cambridge University. That is to say, no matter what type of test and initial state of soil, there is a critical state line in (e, lnp, q) space, which is projected onto the (e, lnp) plane and parallel to the normal consolidation line. The initial state of soil is divided into "wet" and "dry", and shear deformation occurs along this line without volume change. In other words, the concept of critical state is that the volume of soil tends to be constant when the strain is large. The shear resistance and void ratio do not change further, and the strain continues to increase. For normally consolidated soil and over consolidated soil, when the drained shear triaxial compression test is carried out under the same confining pressure, the deformation path is different. But they finally enter the same critical state, which is confirmed by parry. Similarly, if they have the same initial void ratio, the ultimate shear strength will be approximately the same in undrained shear conventional triaxial compression test.

The two basic types and critical states of soil stress–strain curves have been confirmed by a large number of triaxial compression tests. However, there is no clear answer to the mechanism of their production. Therefore, based on a series of drained shear triaxial compression tests of normally consolidated and over consolidated soils, this paper attempts to use the interaction principle between plastic volumetric strain and plastic shear strain (referred to as $\varepsilon_v^p - \varepsilon_s^p$ interaction principle) to clarify the generation mechanism of two types of soil stress–strain curves and the connotation of critical state. That is, the influence mechanism of stress history on the constitutive relationship of clay.

7.2.1 Control of Plastic Volumetric Strain on Stress–strain Curve

According to the principle of $\varepsilon_v^p - \varepsilon_s^p$ interaction, the effect of the plastic body stress ε_v^p on the plastic shear strain ε_s^p is exerted by changing the shear capacity, as shown in Fig. 3.1. In the process of soil deformation, there are two main factors causing volume change: the direct compression of the mean normal stress and the inhibition of the dilatancy. The other is the dilatancy of soil. Under the action of shear strain, the volume can shrink or expand. When the volume shrinks, the internal friction is increased and the shear resistance is increased. On the contrary, volume expansion leads to the decrease of shear resistance. It can be seen from Fig. 3.1 that the change of shear resistance results from the change of plastic volumetric strain. Therefore, the change of plastic volumetric strain directly controls the rise and fall of shear capacity. In this way, if the condition of shear shrinkage is satisfied, the volume shrinkage occurs, the shear resistance increases, and the stress–strain relationship shows strain strengthening. When the dilatancy condition is satisfied, the volume expands and the shear resistance decreases, except for slight shrinkage at the initial stage. The stress–strain curve belongs to strain softening type. In the conventional triaxial compression test with constant confining pressure, the normally consolidated soil satisfies the condition of shear shrinkage. The volume strain increases monotonously during the deformation process. The stress–strain curve is approximately hyperbolic. However, for over consolidated soil, the stress–strain curve shows a strain softening type with a hump. Therefore, it can be concluded that the change of plastic volume strain, that is, the irrecoverable volume deformation, determines the type of stress–strain relationship curve.

7.2.2 Conditions of Dilatancy and Shrinkage

Shear deformation can cause volume change, whether shrinkage or expansion, which is called dilatancy. Reynolds describes the volume expansion observed when granular materials are subjected to shear. The term dilatancy is used for the first time. According to the principle of $\varepsilon_v^p - \varepsilon_s^p$ interaction, dilatancy and shrinkage are the two modes of action of plastic shear strain on plastic volumetric strain, as shown in

Fig. 3.1. The conditions of these two modes of action depend on two factors. One is the compactness of soil itself. The other is the restraining effect of external restraint pressure (mean normal stress p) on dilatancy. In fact, the compactness of soil is also relative to the external constraint pressure. With the increase of constraint pressure, the compactness of soil can be improved.

Casagrande introduced the concept of critical void ratio for the first time in the study of granular soil liquefaction. Under a certain constraint pressure, when the soil shear deformation finally reaches failure, the volume does not change. And the void ratio maintains a constant value, which is called critical void ratio. Similarly, for a specific void ratio, the volume remains constant at failure. At this time, the constraint pressure also remains unchanged, which is called critical constraint pressure. With the introduction of critical void ratio and critical confining pressure, the conditions of dilatancy and shrinkage can be determined quantitatively. A relative constraint pressure parameter p_{ratio} is introduced, $p_{ratio} = P/Pc$, where p is the constraint pressure of the soil at that time and p_c is the critical constraint pressure corresponding to the void ratio of the soil. When $p_{ratio} > 1$, the soil shrinks. When $p_{ratio} < 1$, the soil expands. Similarly, the critical void ratio can be used to judge. In a word, under the action of shear strain, the condition of dilatancy or shrinkage is determined by the void ratio and external constraint pressure.

7.2.3 Critical State

According to the principle of $\varepsilon_v^p - \varepsilon_s^p$ interaction, the interaction between plastic volumetric strain and plastic shear strain always runs through the deformation process of soil before it enters the critical state [11, 12]. The three basic deformation characteristics of soil compression, dilatancy and stress path correlation are the concrete reflection of this interaction. When the soil enters the critical state, the void ratio tends to be constant, that is, to reach the critical void ratio. So that the volume strain ε_v^p will remain unchanged. And the corresponding constraint pressure p also reaches the critical value, then $\mathrm{d}p = 0$. The total strain εv is decomposed into the sum of elastic and plastic volumetric strains.

$$\varepsilon v = \varepsilon_v^e + \varepsilon_v^p \tag{7.1}$$

where, ε_v^e is the elastic strain; ε_v^p is the plastic volumetric strain.

$$\mathrm{d}p = K\mathrm{d}\varepsilon_v^e \tag{7.2}$$

where, K is the bulk modulus of elasticity. $\mathrm{d}\varepsilon_v^e$ is the increment of elastic strain.

Since the total strain εv is constant, there is the following formula.

$$d\varepsilon_v = d\varepsilon_v^e + d\varepsilon_v^p = 0 \tag{7.3}$$

As d$p = 0$, it can be seen from Eq. (7.2),

$$d\varepsilon_{\upsilon}^{e} = 0 \tag{7.4}$$

It is obtained from Eqs. (7.3) and (7.4),

$$d\varepsilon_{\upsilon}^{p} = 0 \tag{7.5}$$

Equation (7.5) shows that the plastic volumetric strain remains unchanged. According to the principle of $\varepsilon_{\upsilon}^{p} - \varepsilon_{s}^{p}$ interaction, the interaction between plastic bulk strain and plastic shear strain will not occur in the critical state. In this way, the correlation of the compressibility, dilatancy and stress path disappears, and the critical state degenerates into a pure shear deformation process in which both the elastic and plastic volumetric strains are constant. Since the stress path dependence disappears, the deformation process has nothing to do with the previous stress history.

According to the conclusion of Sect. 5.1.1, the change of plastic volume strain controls the rise and fall of shear resistance. In this way, when the plastic volume strain remains constant, the shear resistance will not change, that is, the residual shear strength f_q is a constant. It is uniquely determined by the stress state variable p and void ratio e at that time.

$$q_f = f(p, e) \tag{7.6}$$

Equation (7.6) is the expression of critical state line in space e-p-q. In this way, the existence and uniqueness of the critical state line in space are proved theoretically.

7.2.4 Test Work

The preparation, saturation, installation and consolidation of clay samples are shown in Sects. 6.1–6.3. See Table 6.1 for the physical properties of the sample. The conventional triaxial compression tests are divided into two groups. The first group is over consolidated specimen, and the test scheme is shown in Sect. 6.4.2. In order to compare with over consolidated samples, the second group of samples are normal consolidated samples, and the test scheme is shown in Sect. 6.4.1.

Figures 6.4, 6.5, 6.6, 6.7, 6.8 and 6.9 show the comparison of stress–strain curves of clay samples under different consolidation conditions. Among them, Figs. 6.5, 6.6 and 6.7 show the case when the consolidation pressure σ_3 is fixed and OCR changes.

It can be seen from Fig. 6.5, Fig. 6.6 and Fig. 6.7 that under the same confining pressure and σ_3 of 50 kPa, 100 kPa and 150 kPa, respectively, the stress–strain curves of normally consolidated soil (OCR = 1) and overconsolidated soil with different OCR show obvious differences. When the overconsolidation ratio is greater than 8, the relationship between principal stress difference and axial strain has obvious hump. After the peak, the stress–strain relationship decreases and the slope of the curve becomes negative. At the same time, the volume expands continuously and becomes more serious with the increase of OCR, which is called strain softening curve. However, for normally consolidated soils, the stress–strain relationship is a strain hardening curve. The principal stress difference increases monotonically with the axial strain, which is approximately hyperbolic, and the volume shrinks continuously.

The confining pressure p of normally consolidated soil is greater than the critical confining pressure p_c corresponding to its void ratio. According to the condition $p_{ratio} = p/p_c > 1$ given in Sect. 7.2.2, the volume of soil will shrink. According to the analysis conclusion in Sect. 7.2.1, the change of plastic volumetric strain directly controls the rise and fall of shear capacity. Therefore, under this condition, the stress–strain curve of soil is strain enhanced. Similarly, if the current confining pressure of heavily over consolidated soil is less than the critical confining pressure corresponding to its void ratio, the dilatancy occurs and the shear resistance decreases continuously, thus the stress–strain curve presents strain softening type. The above analysis results are fully confirmed by the conventional triaxial compression test stress–strain curves of normally consolidated and over consolidated soils shown in Figs. 6.5 and 6.7. In this deformation process, the plastic shear strain acts on the plastic volume strain through dilatation or shrinkage, which leads to the expansion or shrinkage of the volume, and then the change of the volume strain acts on the shear strain by changing the shear resistance. So it is a $\varepsilon_v^p - \varepsilon_s^p$ interaction process. It should also be pointed out that in this process, the dilatancy forms a bad cycle and accelerates the softening process.

It can also be observed from Figs. 6.5, 6.6 and 6.7 that the final shear stress of both normally consolidated soil and over consolidated soil with different OCR tends to the same value under the same confining pressure. This means that the soil has entered a critical state. According to the analysis in Sect. 7.2.3, the critical state is a pure shear deformation process, in which the interaction between plastic volumetric strain and plastic shear strain disappears completely. The effect of stress history no longer exists, so that the critical state has nothing to do with its previous stress history. The stress–strain curves of different OCR in Figs. 6.5, 6.6 and 6.7 show that both shear stress and volume strain are significantly different, but they tend to the same q value, which confirms the above analysis.

According to the principle of $\varepsilon_v^p - \varepsilon_s^p$ interaction, the change of plastic strain controls the rise and fall of shear resistance in the process of soil deformation, which determines the type of soil stress–strain curve. When the volume shrinks, it is strain strengthened. When the volume expands, it is strain softening.

Theoretical analysis shows that the critical state is actually a pure shear deformation process. There is no interaction between plastic bulk strain and plastic shear strain, and the correlation of compressibility, dilatancy and stress path disappears, so it has nothing to do with the previous stress history.

7.3 Effect of Stress Path on Effective Shear Strength Parameters of Remolded Clay

With the expansion of the scope and scale of soil engineering, underground engineering, slope engineering, deep foundation pit engineering and other stability problems will encounter the influence of stress path on shear strength, especially the influence of shear strength parameters [13, 14]. Many studies have shown that the shear strength and total shear strength parameters are related to the stress path. However, there are different understandings about whether the effective shear strength parameters are related to the stress path. Some test results show that the effective shear strength parameter is independent of the stress path. However, some test results show that it is related to the stress path. It should be noted that these tests are consolidated undrained shear triaxial compression tests. The value of shear strength is affected by the measurement accuracy of pore pressure and the value standard of test failure. The undisturbed soil is affected by its initial state and stress history. Therefore, it is an appropriate way to solve this problem to explore the influence of stress path on the effective shear strength parameters on the premise of excluding other factors.

This section discusses the influence of stress path on the effective shear strength parameters of clay. On the premise that the initial state, stress history, drainage conditions, loading rate, test instruments and failure criteria of the samples are consistent, the drainage shear tests of the same normally consolidated remolded clay under DCTC stress path and DPTC stress path are carried out, and the effective shear strength parameters under the two stress paths are obtained. The test results show that the effective shear strength parameters of remolded clay under different stress paths are quite different, and this is qualitatively analyzed.

7.3.1 Test Work

1. **Sample preparation**

See Sect. 6.1 for the specific preparation process of the sample. From the sample preparation process, there are two standards to control the quality of remolded soil samples. One is moisture content, the other is dry density. Both of them must meet the requirements at the same time before they are qualified. In fact, in the process of sample preparation strictly in accordance with the above standards. After repeated many times, the waste sample rate is also high, the test samples used in this book are all qualified samples. See Table 6.1 for the physical properties of the sample.

2. **Drainage conditions**

The shear strength is affected by the failure criterion of triaxial compression test. The widely used criteria include the maximum principal stress difference (σ_1 −

$\sigma_3)_{max}$ (when the specimen shows strain strengthening, the principal stress difference corresponding to the axial strain of 15% is taken as the standard) and the maximum effective principal stress ratio $(\sigma_1' - \sigma_3')_{max}$. During the drained shear triaxial compression test, the pore water pressure U remains at 0, and the total stress is equal to the effective stress. Therefore, $(\sigma_1 - \sigma_3)_{max}$ and $(\sigma_1' - \sigma_3')_{max}$ will occur at the same time or at the same axial deformation. The strength of the sample is not affected by the two values. In order to eliminate the influence of test failure value standard on shear strength value, drained shear condition is adopted in triaxial compression tests under DCTC and DPTC stress paths in this book.

3. **Triaxial compression test**

The triaxial compression tests were carried out with sj-1a triaxial apparatus produced by Nanjing electric power automation equipment factory. See Sects. 6.2 and 6.3 for the process of saturation, installation and consolidation of remolded clay samples. The consolidated drained shear triaxial compression test was carried out in two groups. DCTC stress path was used in the first group. DPTC stress path was used in the second group. The triaxial compression test schemes under the two stress paths are shown in Table 6.2, and the two effective stress paths in *p-q* coordinate system are shown in Fig. 6.1a, b respectively.

4. **Test results**

The relationship between principal stress difference and axial strain under DCTC and DPTC stress paths is shown in Fig. 6.2a, b respectively, and the relationship between volume strain and axial strain is shown in Fig. 6.3a, b, respectively.

It can be seen from Fig. 6.2a, b that the specimens under DCTC and DPTC stress paths show strain strengthening characteristics, but the shear strength of the specimens under DCTC stress path is much greater than that under DPTC path. It can be seen from Fig. 6.3a, b that the volume strain shows pure shear shrinkage under both DCTC and DPTC stress paths, and its volume shrinkage increases with the increase of consolidation pressure, but the amplitude of the volume strain under DCTC stress path is about twice that under DPTC stress path. The significant difference of stress–strain relationship and shear strength between DCTC and DPTC confirms the strong dependence of constitutive relationship and shear strength on stress path.

It can be seen from the above test process that the initial state, stress history, drainage conditions, loading rate, test instruments and test failure criteria of the same kind of normally consolidated remolded clay under the stress paths of DCTC and DPTC are consistent. The influence of these factors on the shear strength of the remolded clay is eliminated. The difference of stress–strain relationship and the shear strength of remolded clay under these two stress paths can only depend on the influence of stress path.

Table 7.1 Stress values at failure point of specimens under DCTC and DPTC stress paths

Stress path	σ_3 (kPa)	σ_1 (kPa)	p (kPa)	q (kPa)
DCTC ($\sigma_3 = 100$ kPa)	100	295.40	165.13	195.40
DCTC ($\sigma_3 = 200$ kPa)	200	560.59	320.20	360.59
DCTC ($\sigma_3 = 300$ kPa)	300	820.77	473.59	520.77
DCTC ($\sigma_3 = 400$ kPa)	400	1058.38	619.46	658.38
DPTC ($p = 100$ kPa)	55	189.73	99.91	134.73
DPTC ($p = 200$ kPa)	124	351.04	199.68	227.04
DPTC ($p = 300$ kPa)	202	495.37	299.79	293.37
DPTC ($p = 400$ kPa)	275	649.88	399.96	374.88

7.3.2 Treatment and Analysis of Test Results

1. **The stress value at the failure point of the specimen**

As the specimens under DCTC and DPTC stress paths show strain strengthening characteristics, according to the above-mentioned failure criteria of triaxial compression test, the principal stress difference corresponding to the axial strain of 15% is taken as the shear failure criteria of specimens. The stress values of failure points of specimens under DCTC and DPTC stress paths are shown in Table 7.1.

2. **Determination of effective shear strength parameters**

According to the stress value of failure point in Table 7.1, the failure stress circle under DCTC and DPTC stress path is drawn, respectively. And the shear strength envelope is made with the dip angle of effective internal friction angle φ_d. The intercept on the longitudinal axis is the effective cohesion c_d, and the results are shown in Table 7.2.

3. **Qualitative analysis of test results**

It can be seen from Table 7.2 that the effect of stress path on the effective shear strength parameters is significant. The effective internal friction angle under DPTC stress path is about 1/4 lower than that under DCTC stress path. The effective cohesion of the specimen is 0 under DCTC stress path, and it can not be ignored under DPTC stress path.

The effective internal friction angle of the specimen under DPTC stress path is lower than that under DCTC stress path. The reason is that the lateral unloading

Table 7.2 Values of φ_d and c_d under different stress paths

Stress path	φ_d (°)	c_d (°)
DCTC	27.5	0
DPTC	20.3	28.7

caused by the decrease of confining pressure in the process of shear leads to the decrease of soil shear capacity.

The reason for the cohesion of the specimen under DPTC stress path is that the drainage boundary condition is double-sided drainage and the loading rate is slow. The stress path increases by 1σ and decreases by 3σ. The confining pressure drop at failure is about half of the consolidation pressure (Table 7.1). Therefore, there is over consolidation effect in the process of drainage shear, resulting in cohesion.

This section discusses the influence of stress path on the effective shear strength parameters of clay. Under the premise of ensuring that the initial state, stress history, drainage conditions, loading rate, test instruments and failure criteria of the samples are consistent, the drained shear triaxial compression tests of the same kind of normally consolidated remolded clay under DCTC and DPTC stress paths are carried out. According to the test data, the effective shear strength parameters of remolded clay under two stress paths are obtained. The test results show that the effective shear strength parameters of remolded clay under different stress paths are quite different. Qualitative analysis shows that the reason for the lower effective internal friction angle of the specimen under DPTC stress path is that the lateral unloading caused by the decrease of confining pressure in the shear process reduces the shear resistance of the soil. The reason for the cohesion of the specimen under the DPTC stress path is the over consolidation effect in the drainage shear process. After excluding the influence of other related factors on the effective shear strength parameters, these differences are caused by the different direction of stress path.

References

1. Gao Z, Hu D, Zhang Q (1997) Study of constitutive model of soil under complex stress path. J Sichuan Union Univ (Eng Sci Ed) 9(1):50–56
2. Ren Q, Wang J (2005) Numerical method in modeling the elastoplastic constitutive relationship of sand under the path of constant proportion of principle stress. J Huazhong Univ Sci Technol (Urban Sci Ed) 22(3):69–71
3. Duncan JM, Chang CY (1970) Nonlinear analysis of stress and strain in soils. J Soil Mech Found Div 96(SM5):1629–1653
4. Lambe TW (1967) Stress path method. J Soil Mech Found Div 93(118):1195–1217
5. Provest JH (1978) Plasticity theory for soil stress-strain behavior. J Eng Mech Div 104(5):1177–1194
6. Ridley AM, Wray WK (1995) Suction measurements: a review of current theory of practices. In: 1st international conference on unsaturated soils, vol 3, pp 1293–1322
7. Roscoe KH (1968) On the generalised stress-strain behaviour of wet clay. Eng Plast 535–609
8. Shen ZJ (1995) Reduction suction and simplified consolidation theory for expansive soils. Unsaturated Soils 1321–1326
9. Valanis KC (1971) Theory of viscoplasticity without a yield surface. Arch Mech 23(4):517–551
10. Wheeler SJ, Sivakumar V (1995) An elasto-plastic critical state framework for unsaturated soil. Geotechnique 45(1):35–53
11. Yang C, Sheng D, Carter JP et al (2015) Modelling the plastic anisotropy of lower cromer till. Comput Geotech 69:22–37

12. Shively HL (2015) A state dependent constitutive model for rockfill materials. Int J Geomech 15(5):969–970
13. Niemunis A, Krieg S (1996) Viscous behaviour of soil under oedometric conditions. Int J Rock Mech Min Sci Geomech Abstr 33(8):356–370
14. Norris MZ, Zienkiewicz VA (1979) Application of an anisotropic hardening model in the analysis of elasto-plastic deformation of soils. Geotechnique 29(1):1–34

Chapter 8
Content Induction and Research Prospects

8.1 Content Induction

According to the principle of numerical modeling, based on the triaxial tests of expansive soil, sand and clay, this book establishes the numerical model of elastic–plastic constitutive relationship of expansive soil, sand and clay. These models are verified, including the following work.

(1) The numerical model of constitutive relation of expansive soil is established. Based on the engineering background of Lin huaigang earth dam, the physical property test, mineral chemical composition test and triaxial drained and undrained tests under different initial conditions (water content and bulk density) of expansive soil are carried out. Triaxial drainage test includes two different water content and bulk density. Four groups of test curves under confining pressure of $\sigma_3 = 50$ kPa, 100 kPa, 200 kPa and 300 kPa were carried out for each soil sample. Triaxial undrained test also includes two different water content and bulk density. Four groups of test curves under confining pressure of $\sigma_3 = 50$ kPa, 100 kPa, 200 kPa and 300 kPa were carried out for each soil sample. Finally, according to the double yield surface model, RBF neural network is used as the inversion tool to establish the constitutive model of expansive soil directly on the test results. The results show that the numerical model can well simulate the stress–strain relationship of expansive soil samples under different initial conditions, which provides an effective tool for this kind of practical engineering.

(2) Verification of the numerical model of constitutive relationship of expansive soil. The established constitutive model of expansive soil is substituted into the finite element program to calculate the stress–strain relationship of triaxial specimen. The shape and load of the specimen are axisymmetric. The results show that the neural network model can well reflect the stress–strain relationship of the specimen.

(3) The surface of constitutive relation of expansive soil with different water content is drawn. It is found that there are great differences between them

Q. Ren and J. Zhou, *Numerical Modeling of Soil Constitutive Relationship*,
https://doi.org/10.1007/978-981-16-3231-0_8

both in shape and value. In addition, there is an optimal water content for a specific expansive soil, and the strength of expansive soil decreases when the water content is greater than or less than this value. This book quantifies the magnitude of this difference. The important influence of water content on the properties of expansive soil is explained.

(4) The numerical model of sand constitutive relation is established. The triaxial compression test and hydrostatic pressure test of medium dense sand under equal principal stress ratio path, conventional path and equal p path are carried out. The equal principal stress ratio path includes $k = 0.3, 0.4, 0.5, 0.6$ and 0.7. The conventional path includes confining pressure $\sigma_3 = 100$ kPa, 200 kPa, 300 kPa, 400 kPa and 500 kPa. The equivalent p-path includes five paths with mean normal stress $p = 100$ kPa, 200 kPa, 300 kPa, 400 kPa and 500 kPa. Based on the numerical modeling method of expansive soil in Chap. 3, the constitutive model of sand under the path of equal principal stress ratio is established. The results show that the numerical model can well simulate the stress–strain relationship of sand samples under different stress paths, which provides an effective tool for practical engineering.

(5) Verification of numerical model of sand constitutive relationship. The established constitutive model of sand is substituted into the finite element program to calculate the stress–strain relationship of triaxial specimens along the path of equal principal stress ratio. The shape and load of the specimen are axisymmetric. The results show that the neural network model can well reflect the stress–strain relationship of the specimen.

(6) The total stress–strain relationship and yield locus of sand under different stress paths are obtained. According to the test data, the stress–strain curve and yield trajectory of sand in the whole stress field (p, q) under the above three stress paths are drawn. By comparing them, it is found that they reflect the changing trend of stress and strain with different stress paths. It is proved that the stress path has a significant effect on the constitutive relation of rock and soil.

(7) Triaxial compression tests of normally consolidated soils under four stress paths, i.e. enhanced p drained shear (DCTC), equal p drained shear (DPTC), conventional undrained triaxial shear (UCTC) and reduced p drained shear (DRTC), are carried out. Through the comparative study of stress–strain curves under different stress paths, the following conclusions can be drawn. Under the four stress paths, the normally consolidated soil samples show strain strengthening characteristics. In the initial stage, the principal stress difference increases with the increasing rate of axial strain (its stiffness) from DCTC-DPTC-UCTC-DRTC. After yielding, the growth rate of principal stress difference with axial strain (its stiffness) gradually flattens from stress path DCTC-DPTC-UCTC-DRTC. The strength of DCTC is the largest, and that of DPTC is about 1/2 of that of DCTC. The strength under the stress path of UCTC is about 1/2 weak. The strength of DRTC under stress path is about 1/3 of that. That is to say, the strength of the specimen decreases from DCTC-DPTC-UCTC-DRTC. For volume strain, it shows pure shear shrinkage under

DCTC and DPTC stress paths, and its volume shrinkage increases with the increase of consolidation pressure. The volume strain amplitude under DCTC stress path is about twice that under DPTC stress path. However, the volumetric strain shows pure dilatancy under the DRTC stress path, and its volumetric expansion decreases with the increase of consolidation pressure. That is to say, the specimen with higher consolidation pressure has less expansion. These phenomena show that the effect of stress path on the stiffness, strength and volume deformation characteristics of normally consolidated soil is considerable and cannot be ignored.

(8) Conventional triaxial compression tests of over consolidated soil are carried out. Through the comparative study of stress–strain curves under different stress history conditions, the following conclusions can be drawn. In the aspect of the influence of stress history on the soil constitutive relation, the over consolidation ratio is the decisive factor for the volume strain. The volume strain is not sensitive to consolidation pressure. For shear capacity, consolidation pressure is the decisive factor. The effect of over consolidation ratio can not be ignored. The over consolidation ratio determines the strain hardening or strain softening. And it determines the degree of strain softening, but the specimen will eventually reach a unified critical state, with roughly the same residual strength.

(9) The numerical modeling method of geotechnical constitutive relationship is extended to the field of cohesive soil. The elastoplastic constitutive models of clay under DCTC, DPTC, UCTC and DRTC stress paths are established. And the stress–strain relationship in the whole stress field (p, q) is given. It is visualized as a spatial strain surface in the stress field (p, q). In particular, the elastic–plastic constitutive models of clay under DRTC stress path and UCTC stress path are established, which provide practical constitutive equations for soil excavation engineering and corresponding soil engineering under undrained conditions. Through the visualization of stress–strain relationship and numerical simulation, it is shown that the numerical modeling method can more comprehensively describe the deformation characteristics of soil under different stress paths.

(10) By comparing the deformation of normally consolidated soil under four stress paths, it is found that there are significant differences in stress range, peak strain, shape of strain surface and variation trend of volume yield trajectory. The shear yield trajectory is similar. These differences are caused by stress path dependence. The comparison of shear and volume yield trajectories of normally consolidated soils under different stress paths shows that the stress path has a significant effect on the evolution of strain hardening of clay. At the same time, in the traditional modeling method of soil elastoplastic constitutive model, it is not accurate to describe the stress–strain relationship of soil by assuming the fixed form of yield surface. The reason is that the yield surface of soil is significantly different under different stress paths.

8.2 Research Prospects

The research of this book shows that the geotechnical numerical modeling method can truly reflect the stress–strain relationship of different soils under different initial conditions and different stress paths. Compared with the traditional plastic potential theory, it has strong advantages and provides a new and effective way for geotechnical modeling. After all, this work is only in the exploratory stage, and there are still many places to improve and supplement.

(1) In the process of the test, the indoor triaxial test is carried out instead of the field test. The conventional triaxial shear testing machine is used instead of the true triaxial shear testing machine. There is no suction test device in expansive soil test. The maximum value of confining pressure in expansive soil test is only set at 300 kPa, so the deformation characteristics under high confining pressure can not be obtained completely. Moreover, when the confining pressure is increased by 100 kPa in the test, the variation range is too large, resulting in a small data density, which can not accurately reflect the stress–strain relationship of each point in the whole stress field. The accuracy of the drainpipe reading is low in the test. And it all depends on the naked eye reading of the observer, there is a reading error. At the same time, the lag time of drainage reading of expansive soil is relatively long. Limited by the test time, it can not be read after the value is stable, which also causes errors. The above shortcomings lead to incomplete and inaccurate test data. We hope to make improvements in the future when conditions permit, so as to get more real data.

(2) In numerical modeling, both BP neural network and RBF neural network have their own limitations in mathematical model. Due to the complexity of the constitutive relationship of rock and soil, the network structure and input–output function need to be further improved to improve the accuracy and operation speed of the network. A new inversion tool is used for numerical modeling.

(3) In the verification process of the model, the structure of the stiffness matrix dep is asymmetric, which increases the difficulty of solving. For the verification of the numerical model, only the finite element calculation of the sample is carried out. In the future, the symmetrical stiffness matrix can be used for foundation settlement calculation, slope stability calculation, dam crack prediction calculation and other practical engineering calculation.

(4) When using the model of this book for numerical analysis, we can consider combining the model of this book with large commercial numerical calculation software (such as ANSYS, ABAQUS) With its good pre-processing and non-linear computing ability, the numerical modeling method can be better extended to other fields of geotechnical engineering, so as to form an expert system for different geotechnical media to analyze different practical projects. At present, the author is still working on this aspect.

(5) When the stress path changes greatly, the constitutive relationship of soil will change significantly. Therefore, the cyclic loading or unloading path test of clay

should be added in the future to obtain the samples of clay unloading model. The elasto-plastic constitutive relation under the corresponding path is established, so that the constitutive model can be applied to simulate the seepage stress coupling problems under the repeated cyclic loading or unloading path in engineering. For example, the consolidation problems under the conditions of dam unloading and impoundment, foundation pit excavation and so on.

To sum up, although there are still some shortcomings in the present geotechnical numerical modeling method, there are still many shortcomings. There are essential differences between numerical modeling method and plastic potential theory. The numerical modeling method has the advantage that the plastic potential modeling method can not match, and it can accurately reflect the stress–strain relationship of soil under different initial conditions and stress paths. With the continuous improvement and development, geotechnical numerical modeling method will have a broad prospect, and will make greater contribution in the actual geotechnical engineering.

The manufacturer's authorised representative in the EU is Springer Nature Customer Service Centre GmbH, Europaplatz 3, 69115 Heidelberg, Germany. If you have any concerns regarding our products, please contact ProductSafety@springernature.com

Printed and bound by CPI Group (UK) Ltd, Croydon, CR0 4YY
30/09/2026
02225330-0003